D1684964

Drought Preparedness and Response

Second Edition

American Water Works
Association

Manual of Water Supply Practices—M60, Second Edition

Drought Preparedness and Response

Disclaimer

Editorial Manager – Book Products: Melissa Valentine
Technical Editor: Suzanne Snyder
Cover Design: Michael Labruyere
Manuals Specialist: Sue Weikel/Willadee Hitchcock

Library of Congress Cataloging-in-Publication Data
Names: Brown, Chris (Consultant), author. | Maddaus, Lisa A., author. |
   American Water Works Association, issuing body.
Title: Drought preparedness and response / by Chris Brown and Lisa Maddaus.
Other titles: AWWA manual ; M60.
Description: Second edition. | Denver, CO : American Water Works Association,
   [2019] | Series: AWWA manual ; M60
Identifiers: LCCN 2019010087 | ISBN 9781625763334
Subjects: LCSH: Water-supply--United States--Management. | Droughts--United
   States--Management.
Classification: LCC TD223 .D76 2019 | DDC 363.34/929--dc23
LC record available at https://lccn.loc.gov/2019010087

Printed in the United States of America

ISBN   9781625763334   eISBN-13   978-1-61300-505-7

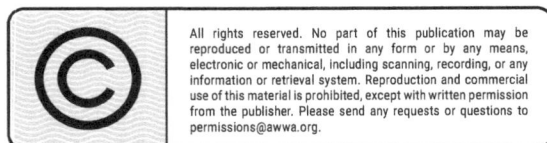

American Water Works Association

American Water Works Association
6666 West Quincy Avenue
Denver, CO 80235-3098
awwa.org

Contents

Figures

This page intentionally blank.

Tables

This page intentionally blank.

Acknowledgments

This second edition of M60, *Drought Preparedness and Response,* was approved by the Water Conservation Division and the Technical and Education Council. Following are the primary authors:

Chris Brown, Chris Brown Consulting, Sacramento, Calif.

Veronica Blette, US Environmental Protection Agency, Washington, D.C.

Toby Goddard, City of Santa Cruz, Calif.

Jessica Seersma, CSU Department of Civil & Environmental, Fort Collins, Colo.

Steve Nebiker, Hydrologics, Chapel Hill, N.C.

Josh Weiss, Hazen and Sawyer, Baltimore, Md.

William Granger, City of Sacramento, Calif.

Lisa Maddaus, Maddaus Water Management, Sacramento, Calif.

Veva Deheza, National Oceanic and Atmospheric Administration, Colo.

Dawn Ison, US Environmental Protection Agency, Cincinnati, Ohio

Nora Mullarkey, Independent consultant, Austin, Tex.

Brian Skeens, Jacobs, Atlanta, Ga.

Our warm thanks to the California Department of Water Resources for its contributions to the first edition, many of which are still evident in the structure and content of this second edition of M60. This edition has been written and revised under the auspices and support of the American Water Works Association's (AWWA's) Planning, Evaluation & Research Committee. In addition to the Committee members listed above, some significant contributions of case study materials from the following individuals helped improve this edition: Drema Gross, Austin Water, City of Austin, Tex.; Jennifer Miles, Regional District of North Okanagan, Coldstream, B.C.; Maureen Hodgins, Water Research Foundation, Denver, Colo.; and Jeff Tejral, Denver Water, Denver, Colo.

This page intentionally blank.

Introduction

The world's supply of drinkable fresh water is under increasing pressure. The United Nations (UN) estimates that water scarcity affects more than 40 percent of the global population and is projected to rise due to effects of climate change and population growth (United Nations 2015). Most people in the United States have easy access to water—it simply comes out of their taps, and it is clean and plentiful. However, increasingly, a growing number of communities are experiencing periodic water shortages. A 2014 report by the US Government Accountability Office (GAO) found that 40 out of 50 state water managers expected shortages in some part of their state under average conditions during the 10 years following their 2013 survey (GAO 2014). Some of the challenges contributing to water shortages today include the following:

- Population growth is a factor, even though citizens may be using less water per person.

- Since the 2011 edition of M60 was published, many areas of the United States have experienced their hottest and driest years on record. According to the National Oceanic and Atmospheric Administration (NOAA 2018), 2012 was the warmest year since 1895, and the four warmest years have occurred since 2012.

- Water is delivered through an increasingly complex and aging network of storage, transmission, and distribution systems.

- Water treatment processes have become more sophisticated and costly.

- Energy-related expenses, from transportation to treatment, have increased significantly.

- The environment is taxed to a critical point in numerous key waterways.

- In dry years, many areas have increased their reliance on groundwater, reducing the future availability of those supplies.

- The reliability of water deliveries has diminished as uncertainty and variability increases, as related to climate change, regulatory actions, delivery system security, and other factors.

There are also new opportunities for reducing the impact of water shortages. Widespread use of the Internet allows for information sharing and communication at a level unimagined in previous decades. New technology allows for more efficient use of water, from commercial cooling towers to smart irrigation controllers. Regional alliances have been established to coordinate water supply and demand management efforts.

M60, *Drought Preparedness and Response*, is designed to help water managers who are facing water shortages. The manual illustrates demonstrated methods of the past as well as new tools and methods. Managing water shortages involves temporarily reducing demand and finding alternate water supplies to temporarily meet demand. Some of these actions will result in permanent changes in water use, such as the installation of efficient toilets. The primary focus of the manual is to provide a step-by-step strategy to anticipate and respond to water shortages through a structured planning process.

AWWA recognizes that the unique aspects of any particular water shortage defy the ability of even the best plans to anticipate and prepare for every contingency. This second edition includes more examples of how water suppliers dealt with multiyear droughts by implementing changes to their programs from year to year or in response to different challenges.

GETTING THE MOST OUT OF THIS MANUAL

Drought and water shortage planning is not just a best management practice for a water supplier; it is a requirement in a growing number of states and water management districts. M60 can help a water supplier meet regulatory requirements for a water shortage plan (WSP)*, but each user will need to consult the rules in his or her state or province where such requirements exist. Typically, such requirements include responding to the drought of record or other predefined challenges.

The manual was written by water professionals experienced with droughts and contains an abundance of information that will help users to write and amend plans to respond to unique and changing circumstances. Finally, while the goal of planning is to anticipate and prepare for future events, experience has shown that the unanticipated can derail a plan's implementation. This manual provides numerous examples of how water suppliers implemented and changed their plans to fit their unique circumstances in instances in which changes to drought measures or stages were required during a drought.

DEFINITION OF A DROUGHT AND WATER SHORTAGES

In the most general sense, *drought* is a deficiency of precipitation over an extended period of time, resulting in a *water shortage* for some activity, group, or environmental purpose. A water shortage occurs when supply is reduced to a level that cannot support existing demands. Natural forces, system component failure or interruption, or regulatory actions may cause these water shortages. Such conditions could last two to three months or extend over many years.

WATER SHORTAGE PLANNING

Providing a reliable supply of water, which is the primary goal of all water suppliers, requires being prepared for water shortages of varying degree and duration.

---

* Water shortage plans are required planning documents in many states and provinces and are known under different titles such as "drought response plan," "drought management plan," "water shortage contingency plan," or "water shortage response plan." For ease of reading, we use the term *water shortage plan*, or the abbreviation *WSP*, in this manual.

Proper planning before a shortage occurs allows for the selection of appropriate responses consistent with the varying severity of shortages. Plans are most effective when water suppliers start demand-reduction measures before a severe shortage develops.

If demand-reduction measures are delayed, reserve supplies may be depleted early in an extended shortage, causing unnecessary social and economic harm to the community. A WSP enables a water supplier to assess the risks and reduce the vulnerability of a community to water shortage impacts and to establish priorities that will provide water for public health and safety and minimize impacts on economic activity, environmental resources, and the region's lifestyle.

DROUGHT-RELATED REGULATIONS AND PLANNING REQUIREMENTS

In many states and provinces, there are regulations that water suppliers must follow when declaring a water shortage emergency that also provide them with authority to enforce emergency measures. Frequently, water suppliers are required to develop and periodically update WSPs as part of their overall water management planning process.

WSPs typically include the following components:

- the policy and legislative intent of the plan, authority for the plan, and public involvement;

- an estimate of supply and demand for five or more consecutive dry years;

- a description of the stages of action to take in response to water shortages;

- a plan for dealing with a catastrophic supply interruption;

- a list of the prohibitions, penalties, and consumption reduction methods used;

- an analysis of expected revenue effects of reduced sales during shortages and proposed measures to overcome those effects; and

- a system to monitor and document water reductions.

SEVEN-STEP PLANNING AND IMPLEMENTATION PROCESS

Water shortage planning is a dynamic process. It evolves as conditions change and new information becomes available. The WSP includes specific mandatory requirements and penalties that become effective when certain shortage conditions or triggers occur. The chapters of this manual describe a seven-step planning process designed to assist water suppliers facing water shortages. Each chapter includes examples and suggestions for communicating the results of the planning step to the public.

The seven steps discussed in this manual are as follows.

Step 1: Form a water shortage response team.
Step 2: Forecast supply in relation to demand.
Step 3: Balance supply and demand and assess mitigation options.
Step 4: Establish triggering levels.
Step 5: Develop a staged demand-reduction program.
Step 6: Adopt the plan.
Step 7: Implement the plan.

Appendix A: Water Shortage Planning Checklist provides an overview of the entire planning cycle as a way to track the tasks in each step. Some of the tasks can be performed simultaneously and are not necessarily in the order that a particular water supplier will follow.

The checklist, combined with the information provided in this manual's seven steps, can help form the foundation of a water supplier's WSP and actions or can be used to update an existing plan.

Appendix B: Additional Sources of Weather and Climate Information provides a list of useful federal and state resources on climate and weather that should be consulted as needed to both prepare and implement the WSP.

REFERENCES

NOAA (National Oceanic and Atmospheric Administration). "Climate at a Glance: National Haywood Plots." https://www.ncdc.noaa.gov/cag/national/haywood (accessed Feb. 11, 2019).

United Nations. Sustainable Development Goals. 2015. "Goal 6: Clean Water and Sanitation." https://www.un.org/sustainabledevelopment/water-and-sanitation/ (accessed Feb. 11, 2019).

US GAO (US Government Accountability Office). *Freshwater: Supply Concerns Continue, and Uncertainties Complicate Planning.* https://www.gao.gov/products/GAO-14-430 (accessed Feb. 11, 2019).

Step **1**

Form a Water Shortage Response Team

SELECTING THE WATER SHORTAGE RESPONSE TEAM

The first step in effective water shortage planning and implementation requires a water supplier to

- designate a water shortage response team leader;
- establish a water shortage response team of staff with decision-making authority representing all departments, clearly defining the roles and responsibilities of each team member; and
- provide the water shortage response team with funding and appropriate level of staff.

Selecting a water shortage response team leader is critical. The designated team leader will spearhead the effort and involve every unit in the organization. The team leader is someone who the water supplier's board of directors and general manager trust to speak for the supplier with the press, lead meetings that hundreds of customers attend, and organize and manage a multiyear shortage response program. This person is someone who is able to work with and motivate all staff and communicate the importance of working together to the staff and community. The team leader should be able to handle several complex situations at one time, deal with the public calmly and consistently, and gain the support of local businesses and groups.

It is the team leader's responsibility to oversee the response team. In selecting team members, management needs to find individuals from numerous disciplines to develop a specific, detailed water shortage plan (WSP). Developing the WSP requires collecting and analyzing water supply and water demand data and understanding the water supplier's sources of supply, operational constraints, critical customers, demand-reduction measures (including estimated savings), community outreach protocols, budget, costs, and sources of income. These elements must be revisited and updated in response to actual conditions, such as when the weather forecasts and water supply conditions indicate the WSP may be implemented within a year.

Every department of the water supplier will be involved in developing the specific WSP and in implementing the plan. For instance, the billing department may need to change the billing format so that customers can compare their monthly water use with the targeted reduction. Meter readers may need to read meters more frequently (i.e., once monthly instead of every two months), and computer programmers may need to develop new account databases to track customer penalty charges and rebate program participation. For suppliers who use advanced metering infrastructure (AMI), increases in the types and frequency of customer reminders and alerts for excessive water use are in order. Rebate programs may need to be started or expanded. Human resources may need to hire temporary staff, and engineering may need to deepen wells, install new water treatment devices, or design system interconnections. Operations may need to accelerate capital improvements or add leakage repair crews as drying soils create more stress on pipes. If the supplier has water conservation staff, they will be essential in many of these actions. However, the implementation of a water shortage response plan will probably affect every staff member's responsibilities. All staff members will be viewed by the community as informal sources of information and should receive sufficient and regular briefings so that they are able to answer basic questions.

Smaller water suppliers may have only one or two people to do all the work. In medium to large organizations, the water shortage response team will probably include more members, some of whom will be more heavily engaged and some of whom will be brought in to work on specific components of the plan or its implementation. The following list shows the types of roles people on the Drought Response Team can play. Larger utilities create teams with many or even all of the following actors.

- Board of Directors or City Council—responsible for approving water shortage–related actions, which often includes adopting a water shortage declaration at various stages of the plan (unless such authority has been delegated to the General Manager), approving implementation of water rate surcharges, adopting updated plans and codes, and approving contracts; a governing body representative may serve as a liaison to a community advisory committee and on a water shortage response team or otherwise be engaged with the General Manager and staff

- General Manager or Director of Utilities—responsible for overall direction for the response

- Water Shortage Response Team Leader—responsible for leadership, management, coordination, information gathering and dissemination, key support staff assignments, role clarification, and communication with a broad array of interested parties

- Water Quality and Treatment Manager—responsible for overall guidance on drinking water quality and operations, issues related to potential alternative supplies, and opportunities for use of nonpotable water

- Water Supply Manager—responsible for estimating and managing long-term water supplies

- Finance Manager—responsible for cost estimates for operational changes, supply alternatives, and demand-reduction measures; customer database improvements and bill format changes; expected lost revenue estimates; recommended rate changes; and use of the revenue stabilization fund

- Conservation Manager—responsible for water use reduction measures management, cost estimates to achieve demand reductions, and liaison with green industry and large water users (residential and commercial, industrial, and institutional customers); this person may serve in the role of Water Shortage Response Team Leader

- Communications Director (wholesale and large retail agencies)—responsible for messaging, customer relations, media relations, press releases, and coordination with wholesale customers

- Planning and Engineering Manager—responsible for new connection water use projections, new and expanded supply infrastructure, interconnection planning, and water quality treatment improvements

- Distribution System Operations Manager—responsible for overseeing frequency and intensity of leakage management, system water loss audits, and meter accuracy testing

- Customer Service—responsible for customer contact, current information about the state of the program, and increasing frequency of customer meter reading

- Administrative Staff—hires staff, purchases equipment, and negotiates union contract adjustments

- Legal Staff—reviews legality of program, rate changes, interagency agreements, and contracts

- Environmental Review (wholesale and large retail agencies)—reviews supplemental supply projects and prepares environmental documentation

---

Tuolumne Utilities District (TUD) Water Shortage Response Team

In 2013, following the August–October Rim Fire in the Stanislaus National Forest and a long-range weather forecast that predicted historically low precipitation, TUD convened its water shortage response team. The team consisted of the General Manager, who was the team leader, the District Engineer, the Water Operations Manager, and the Public Relations Manager and evaluated drought conditions, explored water management and supply options, formulated recommendations to the Board, and implemented all aspects of the drought response.

In addition to ongoing communication, the team met at a designated time weekly. The team brought in other staff as necessary for tactical planning and implementation and provided monthly updates to the Board. The team kept all customer service staff fully and regularly informed of drought conditions, current status, and mandatory conservation measures. All other water supply staff were briefed in case they encountered customers during their daily work duties. Engaging this core team ensured all staff and departments were moving in the same direction, created opportunities for brainstorming innovative solutions, and provided a unified message to the public (USEPA 2018).

Be aware that other governmental agencies, stakeholders, or groups may need to be consulted in developing the WSP and notified when the plan is implemented. For example, the water supplier may need to engage with state or province regulators if plans are required by state law or regulation. Are there other water suppliers or regional authorities with which a WSP may overlap? If the utility received water from a wholesale supplier, a utility representative should be part of the wholesaler's drought response team as well. Are there other local government agencies that may be affected by actions required in the WSP or that may be able to assist in implementing actions? Are there key industries or other large water users (e.g., universities) that should also be engaged in the planning process?

SETTING PRIORITIES

Most water suppliers have a general water shortage response plan that is connected to its water supply or emergency response planning efforts. As these plans are developed or reviewed, it is critical that the utility work with the community in setting priority actions. This will help move towards quick responses when water shortages loom and is critical when facing a drought that may exceed the drought of record. When dry conditions first emerge, it is a signal that the WSP may need to be implemented and should be reviewed carefully and revised as needed to fit the specific conditions of the current drought. The information needing review includes the following.

- Supply: reliability may change from year to year or when triggered by changes in system operational conditions
- Demand: may increase due to unexpected growth or decrease due to efficiency programs
- Revenue: may have changed due to new rates or the depletion of the shortage contingency fund
- Infrastructure: areas in need of repair or subject to pipe breaks may aggravate shortage

Traditionally, water suppliers have focused demand-reduction measures on curtailing customer consumption. Given clear, timely, and specific information on supply conditions and the necessary actions to delay increased reductions, customers prefer the opportunity to meet targeted demand-reduction levels through voluntary measures. The decision to move to mandatory restrictions is more acceptable if the voluntary approach has been tried first but has not resulted in enough demand reduction to ensure public health, safety, and environmental protection through the projected duration of the shortage.

Priorities for use of available water in a community, while varying in order from place to place, may be established as follows:

1. Health and Safety—protect public health and firefighting capabilities
2. Commercial, Industrial, and Institutional—maintain economic base, protect jobs
3. Environment—reduce losses to natural resources
4. Permanent Crops—protect since it takes five to ten years to replace
5. Annual Crops—protect jobs
6. Landscaping—protect jobs, maintain established trees and shrubs
7. New Demand—generally, two years of construction projects are already approved

While reducing customer consumption is important during a water shortage, the water supplier must also focus on the efficiency of its own operations, particularly optimized leakage control. In a fundamental way, leakage—both in the supplier's distribution piping systems and the customer's service connection piping and internal plumbing—are high priorities in a water shortage situation because

- unlike most customer consumption, which runs intermittently, leakage runs continuously and will only increase until detected and abated and
- leakage provides no benefit whatsoever and wastes water and energy resources in extraction, treatment, and delivery.

While it is important for a water supplier to have an ongoing leakage loss reduction program (see M36, *Water Audits and Loss Control Programs*, 4th ed.), it is especially worthwhile for the water supplier to target excessive leakage losses as a priority in a water shortage situation. Traditional customer demand-reduction measures focus on providing customers with programs and knowledge that help them to reduce their use while still allowing them flexibility and choice in how water is used. The amount of public support and cooperation is likely to be greater if actions are equitable; that is, all water users are experiencing a similar service level and degree of involvement or sacrifice. The water supplier should set a positive example by "taking care of its own house" and communicate to the public the work it is doing to manage and minimize water loss and other operational improvements being made to help respond to the water shortage.

Denver Water Drought Response Plan (2016)

In 2002, the Board of Water Commissioners originally adopted a policy stating that Denver Water's goal for drought response is to preserve the quality of public life and economic activity to the extent possible in the face of a water shortage. Denver Water regularly updates its Drought Response Plan (2016), which outlines specific measures designed to maximize available water supplies and minimize water use. Because every drought is different, the Board can adjust and refine drought response measures based on actual conditions.

Denver Water's guidelines are designed to "maintain the health, safety and economic vitality of the community; to avoid adverse impacts to public activity and quality of life for the community; and to consider individual customer needs as much as possible." The water-use restrictions imposed during the 2002–2003 drought indicated that no single "silver bullet" was effective at encouraging all customers to reduce water use. Instead, a "basket of programs"—restrictions, surcharges, enforcement, incentives, and monitoring and evaluation—is recommended to create an overall atmosphere that encourages water savings.

With respect to setting restrictions, Denver Water adheres to the following principles as much as possible.

- Avoid irretrievable loss of natural resources.
- Restrict less essential uses before essential uses.
- Affect individuals in small groups before affecting large groups or the public as a whole, allowing as much public activity as possible to be unaffected.
- Minimize adverse financial effects.
- Implement extensive public information and media relations programs.

ESTABLISHING SCHEDULES AND MAINTAINING MOMENTUM

Depending on the level of preparedness, implementing a WSP will typically require two to six months of advanced planning and dedicated effort. For instance, if rationing is planned to take effect on May 1, the water shortage response team would need to begin work by November 1 of the previous year, or by February or March of the current year at the very latest if there is not ready capability in the existing billing system. Most rationing schemes require custom computer programming to execute.

Step 7 includes a list of essential elements associated with developing the WSP. Table 1-1 contains an example of a planning timeline—any individual timeline will depend on the level of review and activities required by the water supplier to adopt and implement a plan.

COORDINATION, COOPERATION, AND COMMUNICATION

The development of a good WSP is contingent upon effective coordination, cooperation, and communication within the water supplier; with the community; and among local, regional, and state agencies. Following are some strategies for the water shortage response team to consider.

- Establish a community advisory committee. If the water supplier will be asking the public to carry out voluntary water reductions, this group may be able to help develop messages that will resonate with the public. It may also help in building support for water use reductions. Some water suppliers have used a community "ambassador"-type role for these committee members.
- If the water supplier is a city or county, include departments such as parks, fire, health, and the office of emergency services. It will be important for these departments to understand where they place among the priorities for receiving service and how they will be affected under escalating shortage stages. A number of states and provinces, including California, Georgia, Texas, and others, have priority water uses written into code. Water suppliers should check their state or province regulations to be sure their priorities follow applicable law.
- Organize meetings and create partnerships with sanitary districts, local cities, counties, Native American tribes, other water suppliers, regional heath and water quality boards, etc., to facilitate water shortage response coordination.
- Establish a regional public communication program. This is particularly important if the water supplier is in an area with multiple water suppliers and overlapping media markets.
- If others use the same water sources, establish a joint operations liaison between water suppliers or, depending on the complexity, convene a committee to coordinate withdrawals and pumping—quantity and timing.

Effective communication is particularly essential to the success of any WSP in achieving the desired water use reductions. All customers need to be adequately informed about water supply conditions, understand the need to conserve, and know what actions they are being asked or required to take to mitigate the shortage. Even before formal declaration

Table 1-1   Example planning timeline

24 Nov	Staff member reviews water supplier's drought trigger dashboard and the National Oceanic and Atmospheric Administration (NOAA 2019) Drought Outlook forecast and sees potential for initiation of drought conditions within the next 6–8 months.
26 Nov	Staff member begins to research and draft an updated plan.
10 Dec	Draft plan is ready for staff review. General Manager and team review plan and suggest changes. The plan is modified and expanded to include implementation procedures.
17 Dec	Essential staff members review the draft plan, comment on how it affects their functions, and ensure that it is workable.
3 Jan	Board subcommittee reviews the draft plan, suggests changes, and sends the plan to the Board for review and action.
January–March	Public hearings announced. Plan released for public review.
Feb–Mar	Public hearings result in public pressure to revise specific elements of the plan.
End of March	Full Board reviews the draft plan and schedules public hearings.
31 Mar	The Board declares a Water Shortage Warning, requests 10% reduction (*rainy season is almost over*) and adopts Plan.
April	Customers are notified by direct mail that mandatory rationing has been adopted and how the plan will affect them.
mid-Apr	Customers receive individual letter with their allotment, description of rationing plan and appeal procedures, general rationing/information brochure, and conservation information on how to reduce use (efficient toilets, showerheads, and landscaping, meter reading, leak repair, etc.).
1 May	Board declares a Water Shortage Emergency, Stage 1 (*rainy season is over*).

of a water shortage occurs, a public information and media program should be activated to provide customers with as much notice as possible. The more severe the shortage, the more vigorous the public information campaign will need to be. Droughts can be prolonged events that evolve slowly over time. Plan to provide public information using a variety of methods at regular intervals, repeated often and updated frequently. Public information efforts should always strive to be clear, professional, consistent, straightforward, reasoned, and honest to build trust and community support.

Step 7 includes a section that highlights important considerations in developing a public information and media program that should accompany a robust WSP.

REFERENCES

American Water Works Association (AWWA). 2016 (4th ed.). M36, *Water Audits and Loss Control Programs*. Denver: AWWA.

Denver Water. 2016. *Drought Response Plan*. https://www.denverwater.org/sites/default/files/2017-05/DroughtResponsePlan.pdf (accessed Feb. 11, 2019).

NOAA (National Oceanic and Atmospheric Administration). Climate Prediction Center. 2019. "U.S. Seasonal Drought Outlook." http://www.cpc.ncep.noaa.gov/products/expert_assessment/sdo_summary.php (accessed Feb. 11, 2019).

State of California Department of Water Resources. Urban Drought Guidebook 2008 Updated Edition. https://water.ca.gov/LegacyFiles/pubs/planning/urban_drought_guidebook/urban_drought_guidebook_2008.pdf (accessed Nov. 5, 2018).

USEPA (United States Environmental Protection Agency). 2018. *Drought Response and Recovery: A Basic Guide for Water Utilities*. https://www.epa.gov/sites/production/files/2017-10/documents/drought_guide_final_508compliant_october2017.pdf (accessed Feb. 11, 2019).

This page intentionally blank.

Step **2**

Forecast Supply in Relation to Demand

Supply and demand data are needed as a basis for planning and estimating how much water of acceptable quality will be available under various shortage conditions, including multiyear shortages. Pumping and pipeline capacity also are considered in this step. Calculating projected demand, including increases because of growth and less precipitation, will be balanced against projected supply. The best time to initiate this process is before a shortage occurs.

Water shortage planning includes a process of defining possible responses to an array of "what-if?" scenarios. Good planning backed by accurate data produces wise decisions when faced with specific situations.

Information about both historical and current conditions is necessary. Historical data can be used to generate a reasonably accurate description of normal versus drought water supply conditions. Depending upon the state or regional governance, the drought of record or some specific guidance may exist that a water supplier must consider when defining a water shortage. A review of the current supply is used to estimate how much water of acceptable quality will probably be available. Historical and current data, along with projected weather and climate data, are used to create water shortage scenarios. These scenarios account for shortage periods exceeding the drought of record by one or more years. This step reviews the data needed to assess possible water shortage scenarios and provides the calculations necessary for interpreting the data. Figures 2-1 and 2-2 are satellite images of Folsom Reservoir taken before, during, and after the California drought of 2014–2017 that illustrate the effects of drought on a reservoir and the surrounding landscape.

What once was considered a good prediction based on historical data now has a new level of uncertainty. In today's world, water managers can no longer assume *stationarity*. That is, the raw data upon which projections have been based in the past are no longer stationary but vary considerably over time and cannot alone be used to project future

Source: US Geological Survey

Figure 2-1   Folsom Reservoir, California, 2010 (top left) and 2013 (top right) before drought and 2014 (bottom) during drought

conditions. The new reality is the increased variability of precipitation. More suppliers are considering the potential effect of climate change when determining the reliability of their water supply.

Water suppliers will want to incorporate increased uncertainty when implementing the following procedure. To best account for uncertainty, it is essential that planners consider droughts with varying intensities and durations instead of basing planning solely on the drought of record. In order to be aware of drought conditions and forecasts, utilities should routinely check the US Drought Monitor, US Drought Outlook, and other early warning systems.

Source: US Geological Survey

Figure 2-2   Folsom Reservoir, California, 2015 (left) during drought and 2016 (right) near end of drought

DATA COLLECTION

Data concerning permitted water resource allocations, water purchase agreements, available water supply, treatment flexibility, distribution system, customer characteristics, and seasonal demand profiles are compiled and used for building a shortage-planning database.

Weather and Climate Data

General categories of weather and climate information include

- precipitation records and forecasts (rainfall and snowpack),
- stream flow,
- soil moisture, and
- drought indices (e.g., Palmer drought index).

Both historical and forward-looking weather and climate data will be important to review, the former for consideration of the potential impacts of drought and the latter to help inform triggers that will help to determine when a water shortage plan (WSP) needs to be implemented.

A number of federal agencies collaborate to provide a National Integrated Drought Information System (NIDIS), which includes the US Drought Monitor and provides information on the Palmer drought index, weekly and monthly changes in drought conditions, outlooks and forecasts of drought conditions and their impacts, various other water supply and climatological data resources, and regional drought monitoring information. Additional data can be found on the National Weather Service and US Geological Survey websites, and several states have resources specific to their areas. Information about NIDIS and related programs are provided on the following three pages.

Platforms to Help Monitor for Drought

Since the first edition of M60 was released in 2011, additional tools and resources have been released that will help water managers better prepare for drought conditions by monitoring weather and climate. Many of these new tools are available due to improvements in federal interagency collaboration efforts that also bring in state and local government and non-government partners to improve understanding of weather and climate conditions. The **National Integrated Drought Information System (NIDIS)** managed by the National Oceanic and Atmospheric Administration (NOAA) is the primary hub of federal collaboration to provide information to the user community. NIDIS oversees the US Drought Monitor described on the next page. NIDIS has expanded its offerings over the past several years and has many resources that can help decision makers stay on top of the climate outlook for their region.

More recently, NIDIS has established several **Drought Early Warning Systems (DEWS)** across the country. The DEWS use new and existing networks of federal, tribal, state, local, and academic partners to make climate and drought science accessible and useful for decision makers and to improve the capacity of stakeholders to monitor, forecast, plan for, and cope with the impacts of drought. The DEWS offer a wealth of information that is specific to the states and basins in which they operate. Many provide periodic briefings via webinar to provide the user community with current information. The DEWS are very interested in responding to the needs of users in their networks, and there are good opportunities for water utilities to become more engaged. Learn more about the DEWS at https://www.drought.gov/drought/regions/dews.

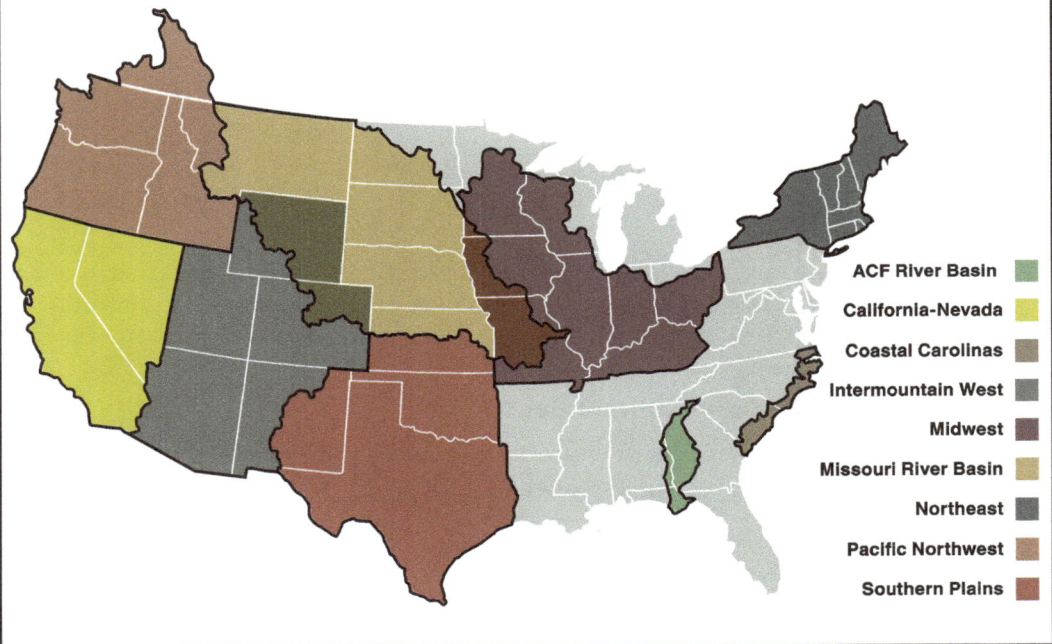

ACF: Apalachicola-Chattahoochee-Flint

U.S. Drought Monitor
www.drought.gov

Background

The US Drought Monitor, established in 1999, is a weekly map of drought conditions produced jointly by the National Oceanic and Atmospheric Administration, the US Department of Agriculture, and the National Drought Mitigation Center at the University of Nebraska–Lincoln. The US Drought Monitor website is hosted and maintained by the drought center.

Process

The weekly maps are released at 8:30 a.m. Eastern Standard Time each Thursday and are assessments of past conditions based on data through 7 a.m. Eastern time (8 a.m. Eastern Daylight Time) the preceding Tuesday. The map is based on measurements of climatic, hydrologic, and soil conditions as well as reported impacts and observations from more than 350 contributors around the country.

March 27, 2018

March 27, 2018

Drought Impact Types:
~ Delineates dominant impacts
S = Short-Term, typically less than
6 months (e.g. agriculture, grasslands)
L = Long-Term, typically greater than
6 months (e.g. hydrology, ecology)
Intensity:
D0 Abnormally Dry
D1 Moderate Drought
D2 Severe Drought
D3 Extreme Drought
D4 Exceptional Drought

Drought Classification

The Drought Monitor summary map identifies general areas of drought and labels them by intensity. D1 is the least intense level and D4 the most intense. Drought is defined as a moisture deficit bad enough to have social, environmental, or economic effects. D0 areas are not in drought but are experiencing abnormally dry conditions that could turn into drought or are recovering from drought but are not yet back to normal.

Other Resources

National Drought Mitigation Center:
http://drought.unl.edu
National Integrated Drought Information System:
https://www.drought.gov/drought
US Department of Agriculture:
https://www.usda.gov
National Oceanic and Atmospheric Administration:
http://www.noaa.gov
The National Drought Mitigation Center's website includes a collection of state drought monitoring and planning resources; click "Info by State" on the "Planning" tab.

General Contact Information

DroughtMonitor@unl.edu
US Drought Monitor
National Drought Mitigation Center
University of Nebraska–Lincoln
P.O. Box 830988
Lincoln, NE 68583-0988
Phone: (402) 472-6707
Fax: (402) 472-2946

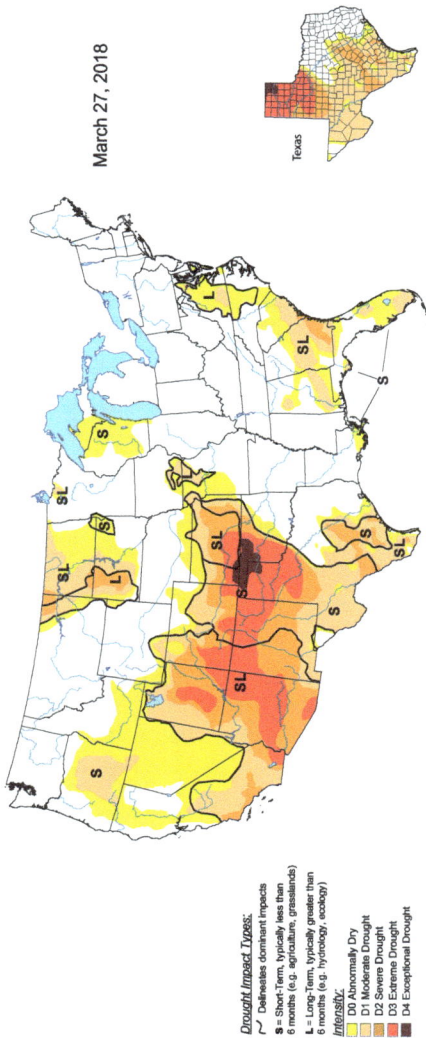

NOAA also oversees the **Regional Climate Centers** (RCCs), another more recent federal-state cooperative effort that is focused on providing climate data, information, and knowledge for decision makers and other users. Many of the RCCs are amplifying and building on tools that have been developed through NIDIS. For example, the Western Regional Climate Center (https://wrcc.dri.edu/) has tools that provide historic atmospheric and precipitation data, current atmospheric and precipitation time series data plots, links to various types of current weather observation data resources, and links to drought monitoring resources and data.

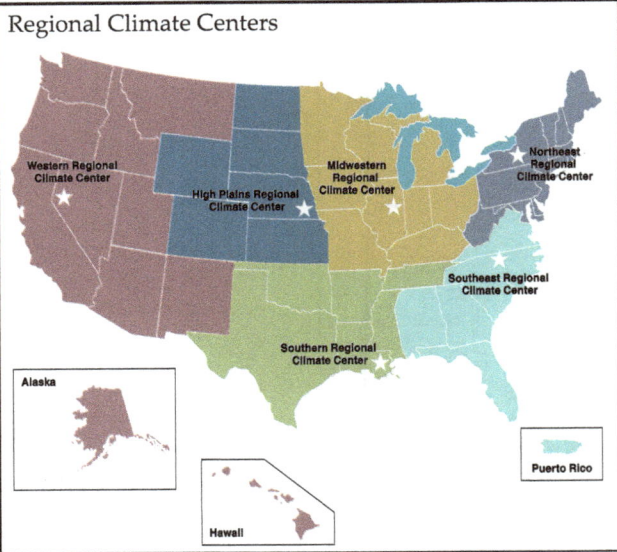

Regional Climate Centers

https://www.ncdc.noaa.gov/customer-support/partnerships/regional-climate-centers

Although not explicitly focused on weather impacts, the **USGS Climate Science Centers** may also offer resources to better understand impacts of drought in an area. Eight regional centers address regional impacts of climate change by offering support and research. https://nccwsc.usgs.gov/

The NOAA **Regional Integrated Sciences and Assessments (RISA)** Program also focuses on research projects that address climate sensitive issues. Some, like the Western Water Assessment (https://wwa.colorado.edu/), are very focused on monitoring drought. https://cpo.noaa.gov/Meet-the-Divisions/Climate-and-Societal-Interactions/RISA

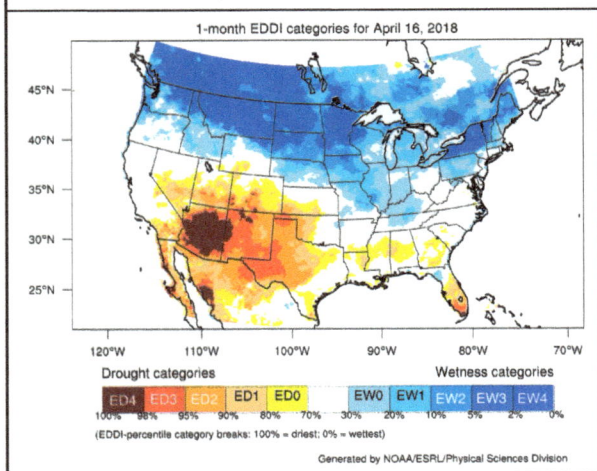

1-month EDDI categories for April 16, 2018

Drought categories

Wetness categories

(EDDI-percentile category breaks: 100% = driest; 0% = wettest)

Generated by NOAA/ESRL/Physical Sciences Division

Emerging Tools

A newer resource that will continue to improve in the next several years is the **Evaporative Demand Drought Index** (EDDI). Still in the experimental stages, EDDI assesses the "thirst of the atmosphere" and can offer early warning of drought and fire-weather risk. Because it captures signals of water stress at weekly to monthly timescales, it may prove to be a strong tool in preparing for flash droughts and ongoing droughts. https://www.esrl.noaa.gov/psd/eddi/

Supply Data

Some of the resources referenced in the previous weather and climate section also maintain information on water supplies, including reservoir levels and groundwater elevations. As geographic information systems are more widely adopted, many states are developing online data dashboards that allow for easy access to water data. For example, the Texas Water Development Board and the California Water Resources department maintain websites (https://waterdatafortexas.org/reservoirs/statewide;   https://water.ca.gov/News/Current-Conditions) that include a wealth of information on the status of supplies in each state.

General categories of supply-related information include the following:

- Supply
  - Local supply status
  - Supply allocation and forecast from water wholesalers and other sources
  - Production records (minimum of five years) and forecasts
  - Surface water elevations and quality by elevation
  - Groundwater table or aquifer levels and quality by elevation
  - Reservoir levels
- Assets
  - Facilities data: maximum sustained pumping rates, pipeline capacities, etc.
  - Contingency agreements for supplemental supplies from regional water agencies or neighboring water suppliers
  - Water rights available for lease

Maintaining detailed records of the state of the infrastructure necessary to store and transport water is essential in accurately determining the amount of water that is available for consumption minus any physical losses due to leakage. This information is also helpful in identifying infrastructure that could be repaired and replaced to reduce leakage and increase supplies. Water audits of the transmission and distribution systems should be routinely performed to ensure accurate calculations of supply and demand.

Supply data should be collected for the past five years or longer, for the drought or droughts of record, and for years when there were diminished supplies due to unforeseen circumstances such as in the aftermath of disasters. The amount of projected water supplies for future years that includes the addition of any new supplies such as the construction of reservoirs, the acquisition of additional water rights, etc., should also be determined.

Regularly review agreements that give the water supplier the ability to provide or receive supplemental supplies during a water shortage. These agreements may need to change if areas experience greater demand due to population growth. If no agreements are in place, determine whether sharing agreements would be beneficial for water suppliers in the region.

Demand Data

Water use data that detail how much water was used by each customer in each sector is needed by month (or as often as is available) for at least the last five years. The amount of water lost due to leakage should be included in the total demand. Projected changes in the population as well as changes in commercial, industrial, and agricultural uses as a result of new development or land use changes should be obtained to determine the future projected demand. Potential increases in water losses due to aging infrastructure should also be considered. These data are used to assess demand for individual customers and customer class and the average use by month, by season, and by year. The following data are needed to assess demand. If the agency does not have access to all of the items on the list, gather what is available, and expand data collection efforts for future potential water shortages.

- Annual volume of water supplied as input to the water distribution system, for at least five years, as well as forecasts for future water delivery
- Billed customer consumption records for at least the five most recent years
- Service area population and growth projections

- Customer class characteristics
  - The average number of residents per single-family residence
  - The average number of residents per multifamily unit
  - The number of units per multifamily connection
  - The number of businesses served by each commercial meter and the number of employees at each business
  - Peak to average residential demand
  - Peak day to winter residential demand
- The number of acres irrigated by each landscape irrigation meter
- The number of acres irrigated by each agricultural meter
- The annual volume of nonrevenue water quantified into apparent (nonphysical) losses and real (physical) losses, which are largely leakage; the cost impact of these losses should also be tabulated
- Attributes of the customer meter population: the number breakdown of meters by age, make, model, and size of meter

DATA ANALYSIS

Much of this information is routinely collected and analyzed as part of periodic updates to water master plans, wastewater master plans, local urban water management plans, and general plan updates or through other agencies for their specific purposes. For example, local business licensing entities may provide the information the water supplier needs regarding the identification of industrial customers through the US Census Bureau's North American Industry Classification System, https://www.census.gov/eos/www/naics/ (2019). Likewise, the county agricultural commissioner, university extension, and resource conservation district may have the irrigated acreage information needed for this analysis.

Coordination of the development of a drought-planning database with ongoing data collection and analysis programs will be beneficial in the long run. If not previously determined in other studies, analysis of the raw data may be required to illuminate certain drought period trends, such as the relation of landscape water use to irrigated acreage and precipitation.

Supply Data Analysis (Projected Dry-Year Supply Without Augmentation)

First, the reliable yield for each source should be determined for the next 5, 10, and 15 or more years. Evaluate what the expected supply would be under various percentage reductions in the overall supplies over varying time periods. For a worst-case scenario, assume a repeat of the worst historical drought in the region or a duration similar to the Millennium Drought in Australia (~13.5 years) to account for longer dryer periods or back-to-back droughts with short wet intervals. Many states and provinces have specific guidelines for the type of analyses. Water suppliers need to check with the agency responsible for regulating water rights or water resource planning in their state for guidance.

For purchased water sources, the contracted allocation should not be used. Instead, the water wholesaler or other source can provide estimates of how much water will likely be available during each of the next five or more years. Active participation in the wholesaler's process is critical to developing the allocation method for short supplies.

Megadroughts and Anticipatory Water Modeling

The possibility and impact of prolonged drought periods—also referred to as megadroughts, which last for two decades or longer and do not allow time to replenish regular and emergency supplies—should be considered. The additional analysis that needs to be performed and the accompanying mitigation and planning strategies for these types of droughts are fundamentally different than the standard drought planning procedures described in this manual. However, it is recommended that water suppliers and government entities go through the exercise of determining the resulting impacts to supply and demand from this type of drought and what the long-term drought management, water supply, water use, and related policies would need to be in order to yield sustainable supplies.

Arizona State University (Gober et al. 2016) evaluated the sustainability of groundwater resources during a 50-year megadrought in the Phoenix area. An anticipatory water policy and planning model was used across a broad range of policies that included population growth management, water conservation, water banking, direct use of reclaimed water, and water augmentation. This analysis not only accounted for weather- or climate-induced megadroughts but also those associated with rapid population growth, overuse, and antiquated management. The analysis showed that continued population growth coupled with no change in per capita demand and current management strategies resulted in unsustainable groundwater supplies even under conditions when there is no megadrought. However, results also showed that moderate reductions in growth along with the implementation of conservation, reuse, banking, and augmentation before the occurrence of extreme drought periods would allow for the achievement of a sustainable supply for many years during and after a megadrought.

Water Quality Data Analysis

Treated water quality by source should be established for constituents regulated by the Safe Drinking Water Act (PL 93-523) and other state, province, or local codes. Comparison of this information with raw water data gives insights into the seasonal variability of the supply and the ability of the treatment system to respond to changes in raw water quality. Data should be developed on how the water quality of each source may vary with the planning cycle use projections. The ability to adequately treat water of degraded quality during drought will be critical if the water supplier plans to use all possible sources. Identify process or chemical changes needed to respond to reduced raw water quality and new supply sources.

Water suppliers that blend water from various sources will need to examine the ability of the treatment system to meet water quality standards when lesser quality water is delivered from one or more sources. Also, certain commercial or industrial customers may need advanced notification if the water quality characteristics will be significantly different during drought periods.

To stay informed of drought-induced water quality issues, a water supplier may decide to temporarily expand its routine water quality–monitoring program. This information may also be useful in alleviating customer concerns if aesthetic differences in water quality, such as chlorides, hardness, and odor, occur during a drought. Increased testing will also provide data on possible impacts of varying water quality on the water supplier and customer distribution systems.

Water Demand Data Analysis (Projected Dry-Year Demand Without Demand-Reduction Programs)

The more information that is known about how customers use water, the better the demand projections and selection of appropriate demand-reduction measures will be. At a minimum, customer type and their seasonal demands should be identified. Standard water supplier customer types include: single-family residential, multifamily residential, commercial, industrial, institutional, landscape, recycled, agricultural, and wholesale.

From a review of water use records, specific water use factors can be determined for each user type on an average monthly, seasonal, and annual basis. Knowledge of user characteristics will be helpful when assessing the demand-reduction impacts of restrictions and rationing allocation methods and estimating revenue generation from pricing changes and water shortage surcharges. Common units of water use measurement are as follows:

- Gallons per capita per day (gpcd)
- Gallons per day per dwelling unit (gpdd)
- Gallons per day per connection (gpdc)

For large landscapes and dedicated irrigation meters:

- Gallons per day per irrigated acre (gpda) by crop, irrigation, and soil type

For some commercial institutional and industrial customers:

- Gallons per day per employee (gpde)
- Gallons per day per square foot (gpdf)

Further analysis of a given category, such as determining individual subarea factors or correlating customer classes with land use designations, may be useful. This is especially important if a large proportion of the overall demand is generated by one or two customer class types.

Water use records should be analyzed on a seasonal basis. Again, conducting this analysis by customer type is essential. Winter residential use compared with summer residential use provides a good indication of indoor versus outdoor usage. The same analysis of seasonal water use among industrial customers may be indicative of process changes or landscape irrigation. Overall, indoor use may be best determined based on inflows, with winter data based on periods when there wasn't any precipitation (to avoid stormwater inflow impacts).

In a dry year, demand for water usually increases over normal demand. More water is needed for landscape irrigation because of less than normal precipitation during the spring and fall, and soil tends to not be as moist. Drought year demand will increase the most in areas such as inland valleys and desert areas that ordinarily use a high percentage of water for landscape irrigation. Some agencies have reported unexpected demand from previously unserved people requesting water service because of failing private wells or for the use of reservoirs for firefighting purposes.

To forecast drought-year demand, it may be helpful to plot normal-year consumption in the service area and overlay that with demand from drier years. Comparing average demand with dry-year demand may provide a multiplier (e.g., 107 percent) to use for predicting the dry-year demand of increased future demand because of growth or other factors. The actual years used will vary based on the region and the quality of data available. Water suppliers may be able to plot full supply periods against data from a historic drought period. Project the dry-year increase for each customer type for future years. If the dry-year demand data is from years where drought restrictions or programs were in place, the dry-year demand information will reflect those, and the multiplier may need to be adjusted to reflect demand in the absence of demand-reduction measures. More rigorous statistical analysis can be performed to identify the factors that impact demand such as rainfall, temperature, season, economy, and the presence of conservation programs. M50, *Water Resources Planning*, and M52, *Water Conservation Programs: A Planning Manual*, provide additional information on water demand analysis and forecasting.

While shortages will vary by year, as population increases and demand grows, additional supplies that could have been accessed during shortage conditions may no longer be available because the water is supplying the new demand.

One concern expressed by some water managers in terms of demand management is commonly referred to as *demand hardening*. This refers to a situation in which most of the waste and low-value uses have been eliminated from the system, so when shortages do occur, their impacts—in terms of damages caused to customers, the environment, and economy—may be greater than in cases where only some of the waste and low-value uses had been eliminated. That is, in a community where most of the toilets and washing machines have been replaced with high-efficiency models and very little water is dedicated to landscape irrigation, there would be less opportunity for significant savings from drought-induced demand management measures. However, in the California drought of 2015 when across-the-board reductions were imposed, even communities with active water conservation programs and diligent water users were able to effectively further reduce per capita water use to very low levels and achieve additional savings. The additional savings were generated by a combination of increased rebate programs for indoor and outdoor water conservation equipment and changes in behavior by customers. Coverage in the media and statewide and local messaging all contributed to the heightened awareness and willingness of customers to change their behaviors. In most communities, demand hardening is not prevalent, and it requires careful analysis of the specific characteristics of the individual system to determine whether or not it is a real problem.

IS THERE A PREDICTED SHORTAGE?

A supplier's projected supply for five or more years, when compared with projected demand for the same period, provides data on the yearly (or monthly) supply and demand balance. Additional scenarios should be developed to account for uncertainty and variability in precipitation. In addition to consideration of the supply and demand over the last five years, scenarios that would occur during a millennium-length drought (as discussed previously) in relation to demand could also be evaluated to determine what the shortages would be under those conditions. Scenarios can also adjust supply and demand for future conditions that include the acquisition of additional supply sources and changes in demand. By adjusting various percentage reductions or increases in supply and demand based upon those scenarios, a supplier can assess what the potential shortages may be. In some cases, these projections may show that the supplies will be adequate for some conditions. It is more likely that during significant dry periods, suppliers will find the supply deficit will vary year to year but gradually increase as reserve supplies dwindle. The predicted, unmitigated annual supply deficit is then countered with supplemental supplies and demand-reduction measures that can be designed for various climate and drought scenarios. A combination of supplemental supplies and reduced demand is used to balance supply and demand.

A *mitigated supply* is the normal dry year supplies plus emergency supplies. *Mitigated customer demand* is the projected dry year demand minus reductions resulting from demand-reduction measures. The combination of emergency supplies and reduced demand is used to balance supply and demand.

Progressive water suppliers are entering into regional, county, and local agreements to improve water supply management, share the cost of emergency supplies, and improve demand-reduction media messages and program costs. The discussion of factors considered in setting drought triggers in Step 4 of this manual is closely related to the evaluation of supply and demand discussed here.

In addition to evaluating and being aware of the supply and demand, it is important to communicate the water supply outlook and the possibility of shortages to the public as conditions evolve. Giving early information on water conditions gets public attention and interest as an early warning before any restrictions may be needed, and it gives the public confidence that there are plans in place to deal with the possibility of a shortage.

A closely coordinated effort between water wholesalers and retailers is essential. Where they exist, a water wholesaler may take the lead and ask their retail customers to share in developing regional and supplier-specific WSPs. However, because many retail customers have multiple sources of supply (wholesale, groundwater, and local surface), they can begin to prepare for water shortages independently. This section discusses how various water suppliers use supply and demand data to guide their response to a supply shortage. In the case where water suppliers have control over their supply systems, they assume the role of both water wholesaler and water retailer.

Actions by Water Wholesalers

Wholesaler water suppliers may manage regional supplies or supplies from primary water suppliers such as state or federal water projects.

Water wholesalers can offer the following types of leadership before and during a water shortage:

- Provide ongoing water supply and demand estimates
- Develop an allocation process with participation of retailers
- Provide retail customers with regular updates to wholesale water supply availability
- Coordinate a consistent regional message
- Coordinate supplemental supply purchases and interconnections
- Coordinate regional demand-reduction strategies
- Coordinate financing for joint supplemental supply and demand-reduction projects
- Coordinate regional or area-wide demand-reduction projects
- Review calibration records for bulk water supply and wholesale meters and launch calibration activities if meters have not been validated recently

Actions by Retail Water Suppliers

Water retailers should make a determination of whether the possibility of a water shortage exists and adopt a current and specific plan meeting their particular circumstances. Managing supply and demand during a drought can be difficult. Consideration of the following items is recommended.

Carryover Storage. If applicable, the water supplier decides how much of the current year supply can be carried over as insurance against a possible subsequent drought year. At a minimum, the carryover amount will be enough to meet essential health, safety, and firefighting needs if the subsequent winters are as dry as the driest years on record. By reducing demand more than is necessary in the first year of a shortage (e.g., 15 percent instead of 5 percent), suppliers have been able to carry over enough supply to avoid increasing reduction targets in subsequent years.

Climate Change. The potential impact of climate change on water resources is another consideration that some water suppliers are exploring. It is expected that one of the impacts of the more variable weather patterns associated with climate change will be longer, drier drought periods. Additionally, climate change may cause many to recalculate the safe yield of their water supply reservoirs because of more severe and prolonged drought periods.

Bay Area Agencies Address Climate Change in Water Plans

The San Francisco Public Utility Commission (SFPUC 2019) incorporates global warming concerns into their forecasting and planning activities. The SFPUC has evaluated the effect of a 1.5-degree Celsius temperature increase between 2000 and 2025 on the Hetch Hetchy watershed at various elevations. It expects that with this rise in temperature, there will be less or no snowpack below 6,500 ft and faster-melting snowpack above 6,500 ft. As such, SFPUC estimates that about 7 percent of the runoff draining into Hetch Hetchy Reservoir will shift from spring and summer seasons to the fall and winter seasons in the Hetch Hetchy Basin. This shift is manageable within SFPUC's planning models, but other water suppliers with storage at lower elevations could be harder hit, especially during longer, drier drought periods.

Similarly, the Santa Clara Valley Water District, the agency that supplies water to much of Silicon Valley, incorporates the possible effects of climate change into its water management plans, much as it already does for earthquakes and flooding. It is partnering with Sustainable Silicon Valley's CO_2 Initiative, a key strategy to respond to climate change by changing the way energy is consumed. The focus is on both energy and water efficiency.

CATASTROPHIC SUPPLY INTERRUPTIONS

WSPs deal with the long-term effects of drought and demand management, but they should also include measures to cope with catastrophic supply interruptions. Commonly, plans are required to prepare for shortages of up to 50 percent. It is important to prepare for both short-term and long-term shortages.

In addition to a WSP, an emergency response plan is recommended. Water suppliers are also encouraged to join their state or province Water/Wastewater Agency Response Network (WARN), which is comprised of other water suppliers offering assistance in times of emergency and accessible on the US Environmental Protection Agency (USEPA) website (2019a). WARN has established a network that effectively puts in place pre-incident mutual aid agreements to facilitate an effective and efficient flow of personnel and resource assistance among water utilities during emergencies. AWWA's M19, *Emergency Planning for Water and Wastewater Utilities* (2018), provides guidance for setting up a WARN process and other steps to prepare for the possibility of catastrophic interruptions of service. Catastrophic supply interruptions have occurred as a result of earthquakes, wildfires, floods, hurricanes, and tornadoes. These catastrophes often create a cascading effect, which is an unforeseen chain of events resulting from the original incident. The Ramona, California, example shows the cascading effect of a wildfire on a drinking water system and community, as well as how beneficial mutual aid can be during an emergency. It is also a reminder that a WSP should have a provision to move immediately to the highest stage of demand cutbacks in case of such unforeseen circumstances.

Cascading Effect of a Wildfire in Ramona, California

During the Southern California wildfires of 2008, in the city of Ramona, a fire-related power failure shut down the local pumping station. When the pumping station stopped sending water to the community, there was no backup generator to take over (voters rejected a proposal to buy one in 1989, so Ramona borrowed three from the San Diego County Water Authority). The lack of pumping caused a pressure loss in the distribution system and resulted in inadequate and nonpotable water being delivered to the community. It took several days for Ramona's drained water supply to be refilled, re-pressurized, and disinfected. Meanwhile, the Ramona Municipal Water District did not have enough employees to service their 10,000 accounts. After some delay, three dozen employees from neighboring water agencies helped restore customer services one by one to avoid straining the system and rupturing pipes.

System interconnections with water suppliers in the region, being a member of the state or province WARN, and participation in comprehensive regional disaster planning can help lessen the effects of catastrophic supply interruptions. In addition to predictable catastrophes such as floods, earthquakes, power outages, and contamination, the physical destruction of facilities as a result of terrorism has taken a higher profile in recent years. The US Environmental Protection Agency (USEPA 2019b) has a variety of resources to help water utilities prepare for and respond to natural and man-made emergencies at https://www.epa.gov/waterresilience.

Many communities now use emergency apps to alert the community of water main breaks, evacuation notices, drinking water contamination, utility outages, fires and floods, and other emergency situations. CodeRED is the most widely used emergency system app of this type and provides alerts tailored to a user's GPS location. It is available for Android, iOS, and Windows smartphones and tablets. For those without smartphones, the CodeRED system also has the capability to send emergency messages via text, voice, email, Facebook, and Twitter. In communities that require residents to opt in or sign up to receive these notifications, it is important to ensure that the community is educated and made aware of the program, which can be done by mailing information about the program and providing information for local schools and employers to distribute to students and employees. Some limitations with the use of CodeRED include periods of data, Internet, and cell phone outages, and areas with limited cell phone coverage. Therefore, it is important that contingency plans are made that allow for the entire community to be informed in lieu of potential data and Internet outages.

CodeRED and Facebook Alert Citizens About Water Emergency

In 2014, Avon Lake, Ohio, experienced an emergency water shortage. All of Lorain county was placed under a state of emergency as a result of the water shortage. The crisis stemmed from ice jamming on the lake that resulted in a substantial slowing of flows entering the Avon Lake treatment plant. This crisis led county officials to send out a CodeRED message request that was also posted to Facebook: "THE CITY OF AVON HAS BEEN NOTIFIED BY THE AVON LAKE WATER PLANT THAT THEY ARE AT A CRITICAL POINT REGARDING THE USE OF WATER. PLEASE DISCONTINUE ANY USE OF WATER UNTIL FURTHER NOTICE." The use of CodeRED has been reported as being effective in relaying emergency information to the residents of Avon Lake.

REFERENCES

AWWA (American Water Works Association). 2018 (5th ed.). M19, *Emergency Planning for Water and Wastewater Utilities*. Denver: AWWA.

Gober, P.; Sampson, D.A.; Quay, R.; White, D.D.; and Chow, W.T. 2016. "Urban Adaptation to Mega-Drought: Anticipatory Water Modeling, Policy, and Planning for the Urban Southwest." *Sustainable Cities and Society*, 27(11):497–504.

San Francisco Public Utilities Commission. "San Francisco Water Power Sewer." https://sfwater.org/ (accessed Feb. 18, 2019).

USEPA (United States Environmental Protection Agency). 2019a. "Mutual Aid and Assistance for Drinking Water and Wastewater Utilities." https://www.epa.gov/waterutilityresponse/mutual-aid-and-assistance-drinking-water-and-wastewater-utilities (accessed Feb. 18, 2019).

USEPA. 2019b. "Drinking Water and Wastewater Resilience." https://www.epa.gov/waterresilience (accessed Feb. 18, 2019).

US Census Bureau. North American Industry Classification System. https://www.census.gov/eos/www/naics/ (accessed Feb. 18, 2019).

This page intentionally blank.

Step **3**

Balance Supply and Demand and Assess Mitigation Options

In case of a predicted or potential shortage, a water supplier has two ways to respond: augment supply and reduce demand. The ability to temporarily augment supply or reduce water demand is specific to each water supplier but in most cases requires careful consideration of numerous factors, including expected severity and duration of the shortage, water quality and other objectives, and cost of mitigation options. This step describes the options available to water suppliers and how to integrate them into a water shortage plan (WSP).

A WSP with an established track record has a strong foundation for effective water shortage management. Even though the emphasis is on water shortages of finite duration, some of the water shortage mitigation measures presented in this step are also appropriate as part of a long-term water conservation and water-loss control strategy. Long-term water conservation and water-loss control measures are valuable to a community in that a given amount of supply can support more users or be available for reserves. However, care must be taken when instituting a WSP concurrently with a long-term water conservation plan. Long-term water conservation efforts can stretch existing supplies, resulting in fewer necessary cutbacks when shortages occur. Water suppliers whose customers are already highly efficient will want to consider that their already existing conservation measures may reduce the number and potential savings attainable through demand management measures during a water shortage.

Water loss control efforts can result in additional demand reduction even if customer demand is already low. Furthermore, long-term conservation efforts may form the foundation for stronger citizen participation and response in times of water shortages. Mitigating against long-term and severe drought conditions requires the use of innovative and comprehensive mitigation measures. The Millennium Drought that occurred in Australia from late 1996 until 2010 serves as an example of the combined effort and fundamental changes required to produce sustainable water supplies for a long-term drought as well as to prepare for future droughts.

Millennium Drought in Australia: Achieving Water and Economic Sustainability

From late 1996 until mid-2010, Australia experienced the longest drought of record that has ever occurred in Australia, referred to as the Millennium Drought or the "Big Dry." This game-changing drought resulted in unparalleled changes to water-related policies and practices, and public perceptions and viewpoints on water conservation, water management, and the value of water in Australia. This drought was all-encompassing and far-reaching, extending to almost all of Australia and impacting millions of people in Australia's most populous cities. In addition to dust storms, wildfires, and loss of plant and animal life, the drought resulted in a $7.4 billion loss from agricultural production as well as the loss of 70,000 jobs in 2002–2003 alone (Australian Treasury 2004).

In addition to the supply augmentation and demand management strategies relevant to droughts and shortages of shorter duration, the Big Dry led to new policies such as

- trading of agricultural water through a market, which involved changes in permitted uses;
- reductions on all water allocations in the Murray-Darling Basin to ensure sustainable supplies;
- government purchase of water rights to sustain the ecosystem;
- increased use of reclaimed water in urban areas;
- investments in desalination;
- imposition of permanent restrictions on water use in urban areas;
- requirements for full-cost pricing on all water uses; and
- incentives for rainwater harvesting.

The water trading has helped to reduce the financial hardships to farmers, contributing $370 million in the Murray-Darling Basin, increasing agricultural production by 2% despite a 14% reduction in water use, and contributing an additional $220 million to the Australian Gross Domestic Product (Heberger 2012). Water trading has been noted as being crucial in mitigating the impacts of future droughts.

SUPPLY AUGMENTATION METHODS

Several options are available for water suppliers to address a near-term supply deficit, representing a range of capital and operating costs, level of control, and applicability for a particular system, scenario, and problem. It should be noted that the kinds of alternatives described in this chapter are not mutually exclusive, and often the optimal solution is a combination of options. An alternatives evaluation should include demand management and operational alternatives in addition to supply and infrastructure alternatives. Screening and feasibility analyses can help to reduce the number of alternatives that are part of the quantitative performance analysis. The supply augmentation options listed in this section are followed by a list of demand management measures for customer implementation in the second part of this step. Methods of supply augmentation can be classified into four groups: (1) leverage existing assets through existing system flexibility and infrastructure upgrades; (2) increase supplier water use efficiency; (3) expand water supply portfolio with new sources; and (4) seek opportunities to collaborate with other agencies. Table 3-1 lists several examples of these methods. Implementation of supply augmentation is often difficult because few of these actions can be undertaken quickly. Also, many of these methods involve balancing environmental and jurisdictional considerations. It is important to consider whether or not the proposed action may be regulated by state or federal environmental laws. Finally, if reserves are available to be used, these supplies must eventually be replenished.

Despite the inherent difficulties with using supply augmentation options, even minimal supply augmentation programs have been helpful in water shortage situations. Developing extra supply increases utility credibility with customers by demonstrating that the water supplier is maximizing its efforts to deal with the water shortage. Also, supply augmentation can provide a water shortage buffer in case of multiyear shortages or can be used to minimize the amount of demand reduction needed to meet temporary supply deficits.

Table 3-1   Supply augmentation methods

Supply Augmentation Method	Examples
Leverage existing assets through existing system flexibility and/or infrastructure upgrades	1. Develop and implement dynamic system operating plans 2. Activate emergency/backup supplies 3. Rehabilitate abandoned sources or infrastructure 4. Increase flexibility of existing assets through infrastructure improvements 5. Blend primary supply with water of lesser quality 6. Transfer surplus water to areas of deficit
Increase supplier water use efficiency	1. Compile an annual system water audit 2. Employ proactive leakage management 3. Reduce distribution system pressure, as feasible 4. Replace inaccurate meters 5. Detect and thwart unauthorized consumption 6. Minimize reservoir spills 7. Suppress reservoir evaporation 8. Recirculate wash water 9. Change billing from bimonthly to monthly 10. Stop turf irrigation at supplier facilities or convert turf to a water-wise landscape and replace inefficient fixtures and irrigation systems at supplier's facilities 11. Require main flushing water capture and reuse 12. Increase system hydraulic monitoring to more quickly identify and respond to failures such as leaks and breaks
Expand water supply portfolio with new sources	1. Increase groundwater pumping 2. Increase use of recycled water 3. Require use of nonpotable water for nonpotable uses 4. Build emergency dams 5. Reactivate abandoned dams 6. Employ desalination—land- or ship-based 7. Import water by truck 8. Rehabilitate operating wells 9. Deepen wells 10. Add wells 11. Renegotiate contractually controlled supplies 12. Use reservoir dead storage
Seek opportunities to collaborate with other agencies	1. Negotiate purchases or options 2. Arrange for exchanges 3. Establish transfers or interconnections 4. Employ mutual aid agreements 5. Develop regional water supply plan

Leverage Existing Assets Using Existing System Flexibility or Through Infrastructure Upgrades

Existing assets can often be leveraged to improve overall performance and squeeze additional water out of the existing supply system. This most often involves modifying standard operations to meet new or changing objectives, often formalized as adaptive management or dynamic system operations. By revisiting established operating rules and protocols, water suppliers can determine whether modified operations could improve the robustness of the system in addressing supply deficits, potentially offsetting or delaying the need for major capital investments.

In evaluating modified operations alternatives to improve reliability, resilience, and sustainability, water suppliers often find that critical limitations prevent them from taking full advantage of system flexibility. This may include limitations in design (e.g., insufficient conveyance or storage capacity); limitations due to infrastructure age or disrepair (e.g., manual or broken valves); and information technology issues (e.g., lack of monitoring and analytical capabilities to support modified operations). To fully leverage existing assets, it may be necessary to supplement the existing system with infrastructure improvements, such as a modified intake structure to allow access to additional reservoir storage or lower river levels or a new pipeline needed to connect a source to new demand areas.

An expanding number of water suppliers are using water supply system models to develop and test alternative operating policies and implement dynamic operating rules that are informed by current and forecasted conditions. New York City's Department of Environmental Protection (DEP) has developed the Operations Support Tool (OST), a daily operations model of their water supply system that is fed by real-time information on reservoir levels, water quality, and other system conditions and driven by probabilistic forecasts from the National Weather Service (NYC DEP 2019). Model simulations with OST help DEP determine the best way to manage the system over the following weeks to months to balance multiple objectives. While the complexity of OST is not necessary for most water suppliers, simulation models can be useful in helping to evaluate the potential for modifying existing operational policies to improve reliability during drought.

Increase Supplier Water Use Efficiency

Water that is conserved through water efficiency improvements represents an environmentally friendly source of supply and is often less expensive. As a standard of practice, a comprehensive supply evaluation should include water efficiency potential and formulate cost-effective demand management alternatives, including conservation and pricing strategies, to evaluate alongside hard infrastructure. In addition, temporary restrictions in water use may be faster and more efficient than permanent augmentation of supply for reliability, depending on economic impacts and willingness to pay.

To win the public's cooperation, water suppliers can demonstrate a visible commitment to efficient water use. Actions undertaken to ensure the efficiency of the water supplier's operating system will save water and set a good example for the public. Suppliers should demonstrate leadership by making a commitment to water efficiency before asking customers to modify their behavior. One example is to reduce or stop turf irrigation and install low-volume irrigation systems for shrubs and trees at all facilities.

Water suppliers should be both *accountable* and *efficient* in their operations at all times—not just in times of water shortage. Historically, most water suppliers have not confirmed their basic level of accountability because they do not compile water system audits (water audits) on a routine basis. In 2016, the American Water Works Association (AWWA) published the fourth edition of M36, *Water Audits and Loss Control Programs*, which provides a standardized, best practice water audit methodology for identifying and quantifying water losses and developed a water audit tool based on M36 that is free to download (AWWA 2016). AWWA recommends that water suppliers compile the standard water audit and water balance annually as a standard business process. Likewise, water suppliers should have ongoing programs to deal with substantial economic losses. It is difficult to launch such programs as a water shortage is unfolding.

As the scarcity of the water supply increases, it is possible that the cost impact of leakage could leap in value and justify significant additional leakage control actions as part of the water shortage strategy. Thus, regular tracking of losses and their cost impacts via the water audit is an essential part of the strategy in responding to a drought or shortage condition. Changing soil conditions with prolonged drought can also lead to increased pipe breakage and leakage and drive home the importance of water loss prevention efforts.

Water suppliers can reduce system pressure to the extent permissible by firefighting standards. A comparison of water-use records of two similar Denver, Colo., neighborhoods indicated that homes with lower water pressure use an average of six percent less water than those with higher pressure. However, the reduction achieved in any system is a function of specific system conditions and results will vary. Fire department pressure checks can be coordinated with main flushing to accomplish both tasks with the same water.

The South Florida Water Management District WSP requires system pressure reductions when there is even a moderate water shortage. Water authorities are asked to reduce pressure to 45 pounds per square inch (psi) at the point of use (i.e., the meter). The water supplier then notifies local fire departments to make arrangements to restore pressure quickly in case of fire. Pressure reduction schemes implemented during emergency conditions should be closely monitored to ensure that necessary water supply is not disrupted. It is recommended to avoid using pressure reduction as a conservation measure during early drought stages because reduced pressure may cause irrigation systems to function poorly.

Water suppliers manage huge quantities of water through expansive water distribution networks that are often suffering deterioration. As a drought or water shortage condition emerges, it is incumbent upon the water supplier to assess the level of accountability and efficiency in their operations—and to take appropriate actions to ensure that undue water waste is eliminated.

Expand Water Supply Portfolio With New Sources

In some cases, it may be necessary to expand an existing water supply through substantial capital investments in new delivery or storage infrastructure and/or new water sources. Such mitigation options involve considerable outlays of capital but can provide significant increases in the overall reliability of a water supply by relieving the stress on existing infrastructure or supplies and adding redundancy to help mitigate potential for failure. A focus on diversification of supply sources can help prepare a water supplier for the wide range of possible challenging conditions that could arise. Evaluation of potential impacts to the environment or to other regional water users is important to ensuring overall sustainability. M50, *Water Resources Planning*, provides additional information on identifying and evaluating alternative new sources.

The best possible solution is to maintain emergency supplies held in reserve. These are often held in local groundwater basins but can also be located in distant water banks. Evaluating the need for supplemental water sources is an activity that should take place well before a water supplier finds itself facing possible shortage conditions. Supplemental water sources can take years to identify and develop.

However, there may be options that can be explored and activated in the near term. Suppliers with surface water supplies may be able to use the reservoir dead storage to the legal minimum pool.

Lowering Reservoir Intake Pipes in Cummings, Georgia

In 2007, the level of Lake Lanier in Georgia reached historic lows and the City of Cumming's intake pipes began to see the light of day for the first time in more than 25 years. The City began the process of installing an emergency bypass pumping system to guarantee uninterrupted water service for the more than 200,000 citizens living in the Forsyth County, Georgia, area. During November 2007, two electric-powered portable pumps along with two high-density polyethylene "straws" were installed in the lake bed to sip Lanier's receding waters. In response to the officials' forecast that lake levels would fall to elevation 1,035 mean sea level by late December 2007, engineering staff recommended that the transplanted silt and sediment that had been deposited on the lake bottom over the past 50 years be removed to ease the flow of water to the City's remaining viable intake pipe. With the older of two intake pipes now completely above the water level, the City made the decision to lower its planned redundant raw water intake facility to elevation 1020 msl which, when completed, became the lowest intake structure on Lake Lanier.

Groundwater wells can often be deepened and the pump-rate increased for limited time periods. Lower-quality groundwater can be blended or special treatment devices installed. In adjudicated or managed basins, it is sometimes possible during emergencies to temporarily increase the annual amount pumped. Well drillers often have waiting lists for their services during water shortages, so planning ahead and reserving time in their schedules can help insure increased groundwater production when it is needed.

It may be possible to attract new recycled water customers during a drought or obtain recycled water from a neighboring water supplier or sanitary district. Because of regulations on recycled water use, landscape irrigation with recycled water is likely more efficient than potable water irrigation use.

During extreme shortages, expensive new water supplies may be the only solution to meet demands. Desalination, brackish water nanofiltration, temporary pipelines, and even water importation by train or truck become affordable. Nanofiltration can also be used to improve the quality of recycled water, increasing the number of possible customers. Water transfers from willing sellers using available pipeline capacity have become significant sources of supplemental water during shortages.

Seek Opportunities to Collaborate With Other Agencies

Collaborative operation of local or regional supplies, or shared vision planning, has been applied successfully to improve regional water supply reliability during low-flow periods. While collaborative water supply planning and operations provides a framework in which costs and risks become shared, often increasing the overall reliability, resilience, and sustainability for the overall combined system, it does come at the cost of independent control over a water supplier's own future. Many water supply managers are understandably hesitant to cede this control unless the benefits can be clearly articulated and quantitatively assessed.

Nevertheless, there are many examples of successful collaboration between regional entities to improve collective water supply reliability, most notably the cooperative management of the Potomac River Basin water supply system by the major Washington, D.C., metropolitan region water suppliers (Fairfax Water, the Washington Suburban Sanitation Commission, and the Army Corps of Engineers Washington Aqueduct Division). In the early 1980s, to address rising demands in the metro area, a study was implemented to evaluate coordinated management of available water sources. The study demonstrated improvement in overall reliability through coordinated supply management, ultimately formalized by the 1981 Water Supply Coordination Agreement and formation of the

Interstate Commission on the Potomac River Basin's Section for Cooperative Water Supply Operations on the Potomac.

Additional, less drastic options for interagency cooperation may be more appropriate for many systems. In these cases, water purchases, transfers, and interconnections may be pursued to supplement local supplies. Interconnections should be closely evaluated as water suppliers have varying distribution system chemistry, which could adversely affect another supplier, even if the interconnection is only temporary. Rules and procedures vary from state to state and from year to year.

DEMAND-REDUCTION METHODS

Demand reduction is typically the most cost-effective and rapid means to respond to a water shortage. Curtailment of water demand is directed at supplier and customer uses that are inefficient, wasteful, often considered discretionary or less essential (such as landscaping), or able to be temporarily reduced or suspended. Because the supplier may mandate certain demand-reduction actions, enforcement mechanisms are needed for maximum effectiveness of those actions. Customer education about the reasons for restrictions is also essential to success.

Demand-reduction measures vary by the severity of shortage and by stage. Stage 1 is usually voluntary and relies on a public information campaign and enforcement of water waste ordinances. Stage 2 can often be managed with a more intensive public information campaign and mandatory restrictions. Stage 3 and Stage 4 most often require customer allocations and severe landscape irrigation restrictions. Plumbing hardware changes can also yield considerable savings. Incentives to assist customers who reduce demand are offered in all stages but increase in scope with the severity of the shortage. M52, *Water Conservation Programs: A Planning Manual*, provides additional information on planning and implementing water conservation programs.

Public Information Campaigns

A public information campaign is the most common way to combat a water shortage. Benefits of public information campaigns include rapid implementation with no direct cost to the customer and raising public awareness of the severity of the water shortage. Water savings from this measure alone range from 5 to 20 percent, depending on the time, money, and effort spent.

Voluntary measures are normally effective only when the public is convinced that a critical water shortage or drought exists. This can be accomplished by letting the public know how many days of supply remain or showing them pictures of near empty reservoirs. Fortunately, the public is already generally aware when drought conditions occur in an area due to regular weather reporting and heightened media attention.

Commonly encouraged conservation actions for various customer types are summarized in this section.

In 2007, the Water Saving Hero campaign (Figure 3-1) was initiated in the San Francisco Bay area and featured a series of ordinary people saving water in catchy ads splashed across billboards throughout the region. A website directed people to their local water districts for more information about incentives that were available to help them save water.

Also in 2007, the Cobb County Water System in Georgia created the Pick 10 Challenge (Figure 3-2), which offered customers ways they could save 10 gallons of water per day and asked them to commit to choosing several and doing them. Because few people realize how much water they use, this can be informative and motivational.

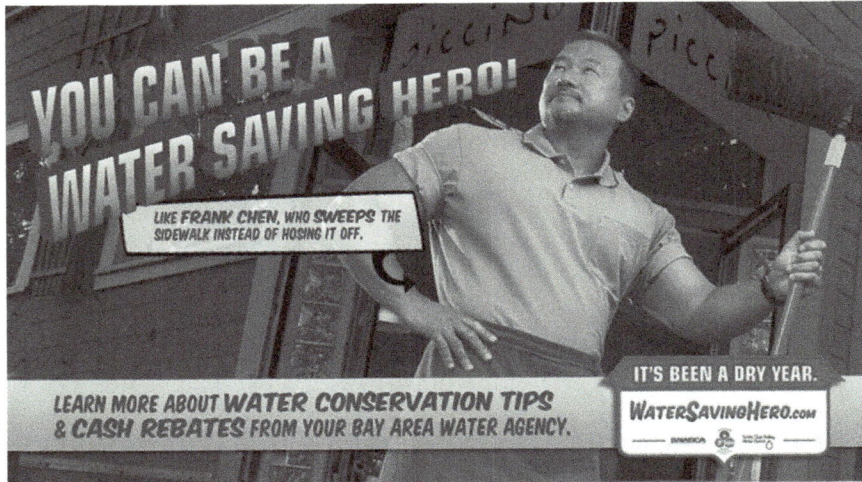

Figure 3-1   Water Saving Hero campaign

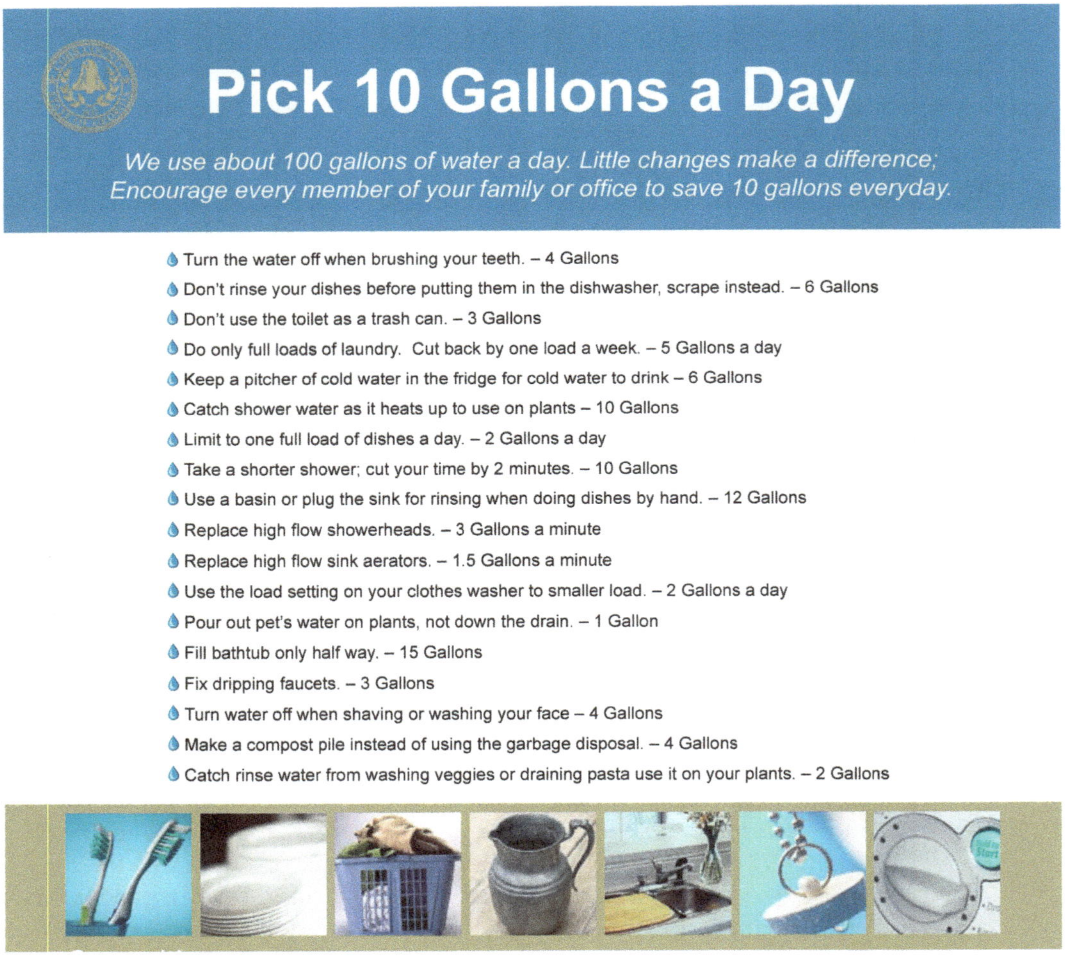

Courtesy of Cobb County Water System, Ga.

Figure 3-2   Cobb County Pick 10 Challenge

Courtesy of Denver Water

Figure 3-3   Denver Water "Use Only What You Need" campaign

Courtesy of City of Sacramento; created by IN Communications and Magma Creative

Figure 3-4   Public awareness campaign

In 2006, Denver Water launched its "Use Only What You Need Campaign" to promote conservation. This campaign resulted in reducing water consumption by about 20 percent between 2006 and 2016 (Figure 3-3).

In 2009, the third year of statewide drought, the California Department of Water Resources in cooperation with the Association of California Water Agencies launched a new statewide public information program, *Save Our Water*. The website, https://saveourwater.com, offers ways to save water through lifestyle and landscape modifications. The *Save Our Water* program was expanded during the recent 2014–2017 California drought using billboards, radio spots, social media, and digital advertising to reach the public.

Sometimes nonprofits, local businesses, or other educational programs will provide or host supportive messaging. The City of Sacramento launched a public awareness advertising campaign to encourage reductions in outdoor water use (Figure 3-4). Independent companies will sometimes donate space on their digital signs or other platforms for the advertisements. It is important to encourage such initiatives.

Voluntary campaigns can also be focused on specific water use, like landscape water use, or directed at target audiences like students. San Diego Water Authority's 20 Gallon Challenge included landscape irrigation and maintenance tips that could save 20 gallons or more per day depending on a homeowner's choices (Figure 3-5). The North Texas Municipal Water District created the Water4Otter program in 2014. It is a program designed to educate youth through in-school presentations (Figure 3-6).

Periods of drought have often been the stimulus for water suppliers to initiate long-term water conservation programs. While the savings from these programs may not be realized quickly enough to mitigate the use of restrictions and enforcement programs, their reliability makes them attractive. Toilet and washing machine retrofit rebates, landscape conversion programs, and evapotranspiration-based irrigation education are all programs that will lead to ongoing savings. These can help a community to avoid restrictions in future water shortages by reducing demand from historic levels. Figure 3-7 depicts a soil amendment program promotion launched by the City of Austin during a drought, the savings from which are based on plant science and future irrigation savings. Separate rebates were offered for compost, mulch, and soil aeration, with an extra amount for customers implementing all three.

Landscape Irrigation

Tip	Estimated Savings
Water only before 6 a.m. and after 8 p.m. to reduce evaporation and interference from wind.	20–25 gallons per day
Don't overwater! 1) Reduce each irrigation cycle by 1–3 minutes or eliminate one irrigation cycle per week. Use the landscape calculator and watering index. Also check out www.sandiego.gov/water/conservation to learn how much to water. 2) Water only after the top inch of soil is dry. 3) Reset irrigation controllers and replace batteries in the spring and fall.	15–25 gallons for each minute; up to 250 gallons per cycle
Adjust sprinklers to prevent overspray and run-off. Repair leaks and broken sprinkler heads. Add 2–3 in. of mulch around trees and plants to reduce evaporation.	15–25 gallons per day 20 gallons per day per leak 20–30 gallons per day per 1,000 sq ft
Install water-efficient drip irrigation system for trees, shrubs, and flowers to get water to the plant's roots more efficiently.	20–25 gallons per day
Upgrade to a "smart irrigation controller" that automatically adjusts watering times for hotter weather and shuts down the system when it rains.	40 gallons per day
Replace a portion of lawn with beautiful native and California-friendly plants. NOTE: These plants do best when planted after winter rains begin.	33–60 gallons per day per 1,000 sq ft depending on climate

Courtesy of San Diego County Water Authority

Figure 3-5   San Diego County Water Authority's Twenty-Gallon Challenge

Courtesy of North Texas Municipal Water District

Figure 3-6   Water4Otter Program

Figure 3-7   Soil amendment rebates offered by the City of Austin, Tex., during a drought

Restrictions

Ordinances banning specific uses of water are forms of mandatory measures. The following are examples of types of demand-reduction ordinances.

Water Waste Ordinances. Ordinances making water waste illegal vary. The following is standard language to consider for a water waste prohibition ordinance.

Waste of Water Prohibited. No water shall be wasted. All water shall be put to reasonable beneficial use. Prohibited water uses include, but are not limited to, the following:

- gutter flooding;
- sprinkler irrigation whose spray pattern hits paved areas;
- use of any ornamental fountain using potable or makeup water for operation;
- car washing except at commercial car washes that recycle water;
- use of potable water from hydrants for nonpotable uses;
- washing of sidewalks, streets, decks, or driveways (except for public health and safety);
- pressure washing of buildings (possible exemption for building rehabilitation projects such as painting); and
- untended hoses without shutoff nozzles.

Landscape Irrigation Ordinances. There are numerous approaches to improving landscape water use efficiency. Four commonly used approaches are as follows.

1. *Watering only between certain hours or on specific days.* If irrigation is allowed less frequently and customers are educated on how much to use, water use has been shown to decrease. Restricting irrigation to twice weekly between specific hours is often recommended during the initial stages of drought. Micro-irrigation of shrubs and trees can be encouraged as this will be a permanent efficiency change. During advanced stages, irrigation can be restricted to once weekly, as it was in Southwest Florida Water Management District (SWFWMD) in early 2009, the third year of drought. By spring 2009, the Tampa City Council issued a complete ban on lawn irrigation with automatic sprinklers that use potable city water. Elsewhere in Tampa Bay Water's service area, SWFWMD imposed what it calls "the toughest water-district restrictions in the state," including limits on the use of automatic sprinklers to one day a week between midnight and 4:00 a.m. In 2008, the service area consumed an average of 253 million gallons daily in April. In 2009, April's daily flow averaged 87.7 percent of that amount.

2. *Watering only with handheld hose or container.* Fixed allocations, allowing customers to responsibly use the water they are allocated as they see fit, allow the water supplier to avoid unpopular water use bans until Phase Four. For example, the North Marin Water District exceeded its rationing goal of 30 percent with its sprinkler ban. It subsequently changed the plan to a voluntary percentage reduction program. A rationing level of 30 percent was achieved through that change, eliminating most of the turf damage that would have occurred if the sprinkler ban had been continued.

3. *Watering only with recycled water.* Lawn watering was prohibited in Corpus Christi, Tex., during a serious drought. The city implemented a program to use recycled water for landscape irrigation and construction uses. Licensed, private tank truck companies delivered recycled water to business and residential customers. Before implementing this program, the public health aspects were addressed by the local public health agency. Regulations set a minimum 1 part per million chlorine residuals to be maintained to all applied reclaimed water. This program was very

successful both for reducing landscape losses and for maintaining jobs and income of severely affected nursery and landscape businesses.

4. *Watering only with graywater.* The use of graywater, untreated household waste water from washing machines, tubs, and showers, is now legal in several states. It is a good way for residents to maintain a portion of their landscapes, especially during dry times.

5. *Watering Restrictions.* In order to more accurately estimate the potential water savings from irrigation restrictions, a water supplier will need to estimate the monthly amount of irrigation use by customer class. Some suppliers have found that multifamily residential accounts use 15 percent or less of their total yearly demand for irrigation while landscape irrigation may account for more than 50 percent of single-family residential water use. The nature of the restrictions used will depend on the severity and timing of the situation. The following are possible water restrictions.

- Prohibit irrigation during the warmest hours of the day, e.g., between 10 a.m. and 6 p.m. Consider allowing irrigation only during early morning and evening when customers can observe the efficiency of the irrigation system. Allowing irrigation between 9 p.m. and 5 a.m. may result in sprinklers running all night or systems with leaks operating for days or weeks before being noticed and repaired. Specify how much time per zone is allowed; during more severe water shortages or when drought impact increase rapidly, these times can be reduced (e.g., reduce watering time from 30 to 15 minutes per zone) without changing the schedule.

- Limit all sprinkler irrigation to a specific number of days per week. The water supplier should also have the authority to limit the amount of water used (e.g., no more than one hour per zone) if water shortage conditions worsen. The number of days will depend on target consumption goals, the time of year and the extent to which irrigation is occurring, and how much demand has already decreased. For example, if demand has already been reduced by 15 percent through other measures, limiting sprinkler irrigation during July and August to *two* days a week could further reduce average daily demand by as much as 5 percent. Limiting lawn or turf watering to one day a week could reduce average daily demand by as much 10 percent but may result in dead lawns. Coupling these limitations with education or public awareness campaigns encouraging people to only use what they need as well as information on the average amount of water they should be applying on that day should help to curtail overuse of water.

Sample schedules include the following.

- Twice weekly (avoid allowing irrigation on weekend days to maximize reduction)
  - For residential addresses ending in odd numbers: Monday and Thursday
  - For residential addresses ending in even numbers: Tuesday and Friday
  - For commercial accounts: Monday and Friday
  - Wednesday, Saturday, and Sunday: no outdoor irrigation with sprinklers
- Once Weekly
  - For addresses ending with
    - 0 or 1: Monday
    - 2 or 3: Tuesday
    - 4 or 5: Wednesday
    - 6 or 7: Thursday

      o 8 or 9: Friday

      o Saturday and Sunday: No irrigation with sprinklers

   – Ban sprinkler irrigation, with low-volume irrigation prohibited during the warmest hours of the day, e.g., between 10 a.m. and 6 p.m.

Savings From Mandatory Measures Far Exceed Voluntary Water Restrictions in Los Angeles

The impact of water conservation and water restriction programs in Los Angeles during a period from 2008 to 2010 on the amount of residential water use showed that restrictions resulted in more savings than voluntary measures (Mini et al. 2015). The researchers found that with voluntary restrictions, there was not a significant reduction in residential water use the following summer. The initial implementation of mandatory restrictions resulted in decreases ranging from 4% to 15% during summer periods. However, with more stringent mandatory restrictions, the citywide residential water use decreased by 23%. The stringent mandatory restrictions involved limiting lawn irrigation to twice per week with an accompanying increase in price per unit in the Tier 2 rate.

As demonstrated in Figure 3-8, irrigation restrictions led to water savings in Austin, Tex. One-day-a-week watering restrictions led to greater water savings than twice a week. With modifications based upon its customer mix, the city was able to achieve a flatter pumping profile in 2013. The specific measures in the final irrigation restrictions were different based on customer type. Figure 3-9 shows both the restrictions and the use of

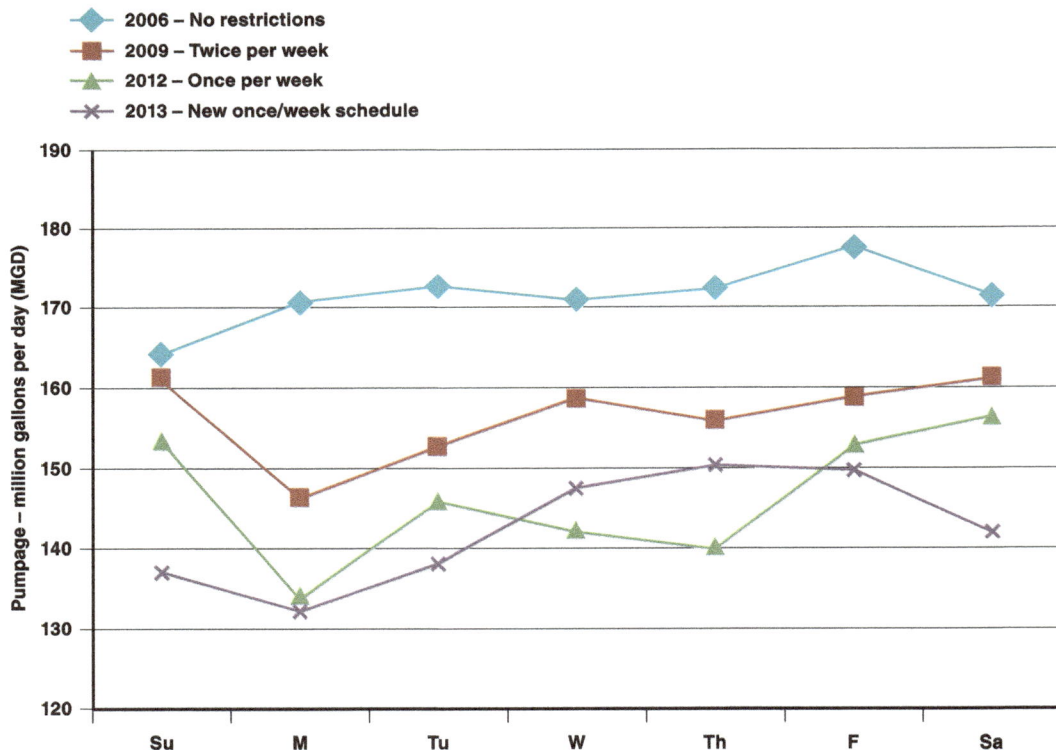

Adapted with permission from City of Austin

Figure 3-8   Demand response to different restrictions—Austin, Tex.

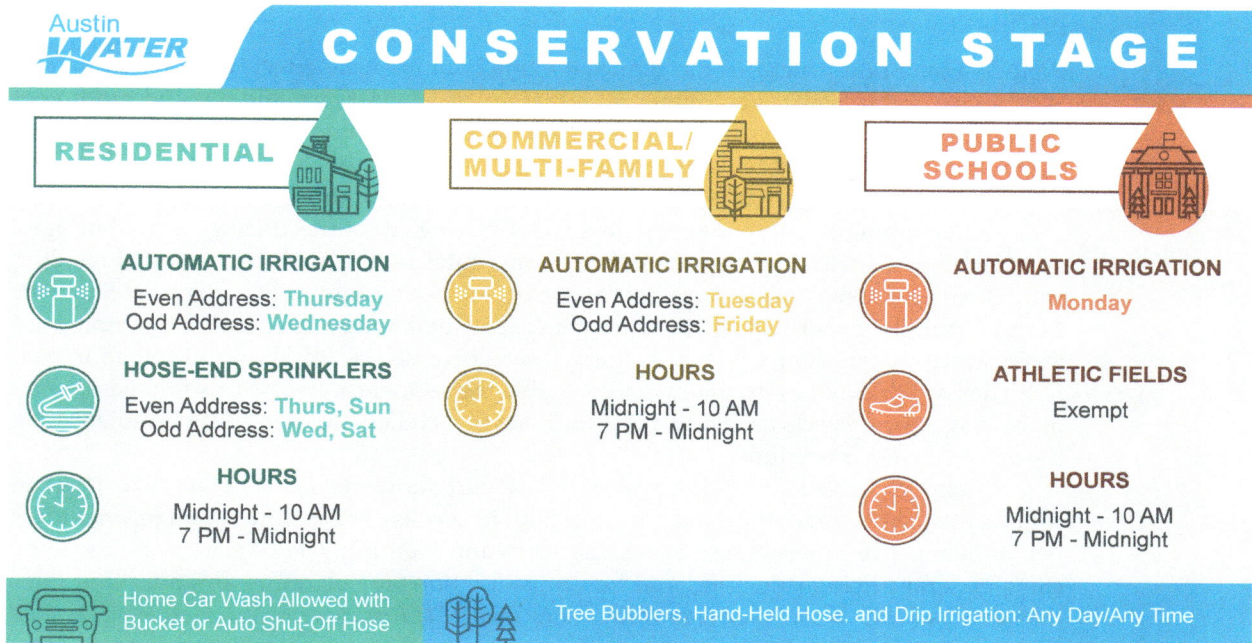

Source: City of Austin

Figure 3-9   Different irrigation restrictions based on customer type in Austin, Tex., 2013

attractive graphics to communicate the different expectations for residential customers, commercial customers, and schools.

Restrictions on landscape water use can be offset with landscape water conservation opportunities and incentives. Offer solutions to permanently shift irrigation practices and address urban landscape and irrigation designs that are prone to water waste. A water shortage provides an opportunity for suppliers to promote low-volume irrigation and turf removal as options for people to consider. These have the advantage of returning long-term savings, although the programs should be designed well before a shortage begins.

Possible Exemptions From Water Use Restrictions

It may be necessary to allow some exemptions to the water use restrictions. Categories of possible exemptions include new landscapes with low-volume irrigation, parks, sports fields, and golf course greens, and the efficient use of nonpotable water for street cleaning, dust control at construction sites, and other nonpotable uses are unlikely to be restricted. Exemptions do not necessarily mean allowing uninhibited water use, but instead can include mandatory measures that allow for alternative patterns. These may make it possible for the use and functioning of commercial and other revenue-generating properties that would otherwise be closed due to mandatory restrictions. Exemptions can be designed to reduce water consumption by, for example, allowing the use of drip irrigation or reclaimed water in place of sprinkler systems or requiring that a certain percentage of irrigation be performed using these alternative irrigation methods. Exemptions can also be structured to vary depending upon the stage of the drought. A strong public education campaign is necessary when allowing exemptions, as customers need to know why certain areas are allowed to be irrigated when their properties are not.

Pricing

Well-designed rate structures can reduce the potential financial effects of water shortages. Surcharges on excessive uses, refinements that reward conservers, and use and funding of emergency abatement measures are all considerations. When revising a normal fee schedule to account for drought, it is important to anticipate possible legal hurdles and state or province laws that may disallow, prohibit, or limit the use of different rate structures.

Water suppliers can implement new water pricing structures during water shortages. In metered areas with volumetric pricing, raising rates on the quantity used will result in water use reductions. A water supplier can expect rapid and significant water use reductions to result from large per-billing-unit price increases. Combining a large per-billing-unit price increase with significant excess use charges can increase the likelihood that the targeted reduction is achieved. Water rates should enable the supplier to recover its purchase, treatment, and delivery costs as well as the additional costs related to the WSP and replenishing the drought emergency fund.

Pricing changes should be a part of a WSP and adopted as part of the plan. This can reduce the rate change approval from months to weeks. The following pages include a list of the most common types of rate structures and their impact on drought followed by some general pricing principles and key considerations for planning and adopting rates that will assist in water shortage mitigation. (See M1, *Principles of Water Rates, Fees, and Charges,* for a more in-depth discussion.)

Inclining Block Rate. The billing rate increases as water use increases under an inclining block rate structure. This encourages customers to save water, and frugal water users will benefit from lowered rates. Include a lifeline rate in the pricing system that is as low as possible for basic health and safety uses.

Seasonal Rates. For seasonal rates, low water charges cover the water production costs in winter; in summer or other peak periods, the rates increase to meet the capital costs associated with the expanded facilities necessary to produce peak demand capacity. These increased summer rates influence customers to reduce water use to lower their costly summer water bills.

Uniform Rate. The same rate is charged for each billing unit consumed. While this method provides some incentive to reduce consumption, it represents a passive rate structure that is not likely to reduce water sufficiently during a drought.

Drought Surcharge. During extreme water shortages, water utilities often institute surcharges to alleviate falling revenues because of decreased water sales. It should be emphasized that these surcharges are separate from normal billing and will be eliminated when the declared shortage ends.

Excess-Use Charge. The excess-use charge is assessed during rationing periods to those customers exceeding their allotments.

Denver Establishes Drought Pricing Principles

There is a relationship between price and demand. In theory, customers respond to an increase in price by reducing demand. The question is at what price level will the customer respond? The answer varies based on a number of factors.

Surcharges will be incorporated into an overall program to increase customer awareness of the drought's severity and the importance of saving water. Customers respond to the "basket of programs" concept, which includes surcharges. Drought pricing plays a role in creating an environment in which customers recognize the importance of reducing water use.

Surcharges may apply to current water demands, new taps, or other demands on the water supply. There is concern about issuing new taps when existing customers are subject to surcharges. Applying various forms of surcharges to different types of demands on the water supply provides an equitable method allowing all customers to share the burden of the reduced supply.

Denver Establishes Drought Pricing Principles

Surcharges are less effective by themselves. Industry studies and Denver Water's own customer surveys indicate that surcharges are more effective at reducing water use when combined with other restrictions to create an atmosphere that promotes water savings. Customer response to price signals varies depending on several factors such as affluence, billing frequency, and the normal cost of water. Empirical data show that customers respond to temporary water pricing strategies as part of a water savings environment.

Surcharges are separate from rates. Rates are based on cost. They are established to recover particular kinds of costs specified by the Denver City Charter. The purpose of drought surcharges, on the other hand, is to raise awareness of the value of water, to reduce water use, and to penalize those who do not comply with drought restrictions. These goals are better accomplished when surcharges are implemented as a temporary measure outside the cost-of-service rate structure.

Surcharges should match the severity of the drought. Because every drought is different, each one may require a different set of responses. Surcharges must be structured to help create an atmosphere of appropriate water savings.

Surcharges must be feasible for computer systems to handle. Denver Water must be able to respond to drought conditions quickly and efficiently. Any change in water use charges must be manageable with only moderate modifications to existing computer systems. Substantial changes increase response times and contribute to errors. Because internal coordination is critical, staff members from Customer Care, Information Technology, Accounting, and other relevant sections will be included in discussions of surcharge options.

Surcharges should be tailored for different customer groups and monitored for effect. A one-size surcharge does not fit all. Commercial and industrial customers use water differently from residential customers. Large-volume public use customers may need some accommodation. The surcharge structure must be flexible enough to promote water savings while still addressing diverse customer needs.

Surcharges should reflect overall drought response philosophies. Because all surcharge structures divide customers into groups, no surcharge structure is 100 percent "fair." Some customers may pay a surcharge even if they comply with the other restrictions. In addition to raising awareness of the value of water and encouraging temporary reductions in use, surcharges can supplement revenues if necessary.

Surcharges may need to be seasonally adjusted. In Colorado's semi-arid climate, water use is greater in summer than in winter. Outdoor use is more discretionary than indoor use, and surcharges should be adjusted to assist in maintaining a water savings environment. Because restrictions to reduce indoor use are difficult to design, adjusting surcharge thresholds can be more effective at monitoring and reducing indoor water demand in winter.

Public input and information are key to customer understanding of surcharges. When surcharges are designed and implemented, the public must have adequate opportunities for input, the surcharge must help create an atmosphere of water savings, and the public must receive adequate information to fully understand the surcharge program.

Surcharges are temporary measures. The criteria that determine when surcharges will be lifted should be specified before the surcharges are imposed. This will reinforce the temporary nature of the surcharge in the minds of customers.

Equity issues related to removing the surcharge should be addressed in advance. The specified conditions leading to termination of surcharges do not always occur at the end of a billing period. Criteria for lifting the surcharge once the specified conditions occur should be considered ahead of time.

Some water suppliers have programs to audit specific water uses. Recommendations are made on how much water is needed for the uses after reasonable conservation measures are implemented. The use of landscape water budgets, where a certain amount of water is allowed per square foot of landscape, is a good example. When rationing is required, these conservation-based amounts can provide the volume of water for the first tier. Water used beyond that can be priced higher to encourage conservation.

Whenever price structure changes are contemplated for use as part of a WSP, a realistic assessment of the length of time to get it approved must be made. Often, utility rate setting takes several months. However, declaring a water shortage emergency allows agencies to make timelier rate adjustments.

Also, it may be unrealistic to expect the conservation benefits of price changes to make an immediate impact, especially if billing cycles are staggered or are bimonthly. However, significant excess-use charges, even where billing cycles are staggered or on a bimonthly basis, can have an immediate and significant impact on demand.

It is standard practice for water suppliers to maintain a dry-year contingency reserve fund to protect revenue through two or more consecutive years of supply reductions below normal demand levels. Rate hikes, surcharges, or borrowing strategies are expected in water suppliers without an established reserve or when the reserve has become depleted.

Unmetered Suppliers and Pricing. Unmetered areas face special challenges implementing drought conservation programs because they cannot impose per customer reductions or per capita allotments. Conservation programs consist of informational programs, restrictions, voluntary measures using rebates and incentives, and technical assistance.

Rationing and Allocation Approaches

In situations when a water shortage emergency becomes so extreme that voluntary reductions, public information campaigns, restrictions, prohibitions, or pricing strategies, either alone or in combination, are unable to produce the needed demand reduction, suppliers may need to temporarily allocate or ration their available supply. Many considerations are involved, including administration, utility billing/data processing capabilities, customer groups affected, enforcement approach, and effect on utility revenue. Water agencies must decide for themselves whether this or some alternative approach is most appropriate for their communities.

Water rationing is different from other use restrictions. The general concept is that each utility customer is given a certain allocation of water to use in a billing period. If they exceed their allocation, the portion in excess of their allocation is charged a penalty rate. In effect, rationing is a strong pricing system to motivate customers to modify their usage so that they stay within their set allocation. Most customers are able to adapt and succeed, particularly when rationing is coupled with a strong information and education component.

Rationing programs often reflect two general approaches: *supplier-oriented* or *customer-oriented*.

- The supplier-oriented approach, where a water supplier develops an overall water use reduction target and applies it to all its customers without regard to individual circumstances, requires little or no additional staff or computer programming to develop.

- The customer-oriented approach requires additional staff and computer programming to develop individualized allocations and more customer service. These programs usually produce significant and sustained demand reductions.

Customer response to rationing and allocation programs is best when the rationale is transparent and the cutbacks are perceived as fair. As an example, Table 3-2 shows the broad range of conservation savings achieved by selected California water suppliers during the 2014–2017 drought and their results. In nearly every instance where mandatory rationing was implemented, consumers responded by reducing water use more than was requested. In 2015, the state mandated water use reduction goals for consumers and water agencies establishing a target of 25 percent total water use reduction for the state. Each water district was assigned a mandatory percentage reduction that needed to be met by its customers, which was determined by the water supplier's July 2014 to September 2014

Table 3-2   A sample of supplier targets and actual savings during California drought, 2014–2017

Supplier	State-mandated Conservation Reduction Targets (February 2016), %	Actual Savings since June 2015 (as compared to 2013), %
Marin Municipal Water District	24	33.3
East Bay Municipal Utility District	16	23.6
Contra Costa Water District	28	34.6
San Francisco Public Utilities Commission	8	14.9
Los Angeles Department of Water and Power	16	15.7
City of Sunnyvale	16	28.1
City of Santa Clara	16	21.8
City of Pleasanton	24	38.7
Cambria Community Services District	8	35

Source: https://www.waterboards.ca.gov/water_issues/programs/conservation_portal/conservation_reporting.html

daily per capita residential water use rate. The rate was periodically adjusted to compensate for water suppliers that were not meeting conservation goals and for those that were exceeding water conservation goals. The cumulative savings reflects the cumulative water use savings from June 2015 through February 2016 as compared with 2013 water use values. A strict rationing program, excessive use surcharges, and the installation of water-saving devices reduced water use and demand. In the East Bay Municipal District, the water demand in 2016 was the lowest it has been since 1978.

During the drought of 2007–2009 in Georgia, the governor mandated that water utilities in the northern half of the state and their customers reduce their summertime use 10 percent below the previous year's winter average. Full mandatory outdoor water use restrictions were already in effect. Through the combination of public information and strict enforcement, water use was reduced by 18 percent.

One of the inherent problems with a rationing system is in accurately designing the program to achieve the desired demand-reduction level without greatly exceeding this amount. Although midcourse corrections can be made to lessen the impact of a program that is too severe, such adjustments are risky and most managers are reluctant to make them. Changing programs too often may send a message to customers that the supplier's planning was faulty. Therefore, it is necessary that rationing and allocation program corrections be presented carefully to customers.

Key elements of a successful rationing program are that the available water is shared as equitably as possible, and that customers are kept informed about the status of the shortage. Allocation disagreements, however, are to be expected, and procedures to handle exceptions and variances need to be part of any well-designed program.

A good public information program helps to administer and enforce a rationing plan. Information regarding water use should be published and supplied at least weekly to keep customers committed. Also, providing fixture replacement rebates, customer water onsite assistance, and useful information to help customers reduce water consumption stimulates relatively painless short-term and long-term water demand reductions.

Rationing programs are generally patterned after one of four basic allocation schemes: (1) percent reduction, (2) per connection allotment, (3) per capita allotment, and (4) hybrid per capita–percentage. Percent reductions can be applied to all customers. The other schemes are only for residential customers.

A *percentage reduction* assigns each customer a consumption reduction goal as a percentage of the consumption level used in a previous year or a five-year average. Required percent reductions can be constant, stepped, or variable. Fixed percentage reductions were used widely during the 1977 California drought. The cities of Concord, Palo Alto, San Mateo, Napa, and Vallejo all used allotment programs that depended on a customer's previous year water use. In Southern California, people were given a baseline allotment of 90 percent of their previous year's consumption with excess use charges for water consumption above that level. The fixed percentage system was easy to coordinate because water allocations were quickly determined from the previous year's water bills.

The percentage reduction method, however, was widely perceived as inequitable because it had the effect of penalizing former water conservers while rewarding those who had previously used large water quantities. Neighbors living in identical houses could therefore receive vastly different water allotments. Also, this plan does not distinguish between indoor and outdoor water use.

During severe shortages, a rationing plan based on percentage reductions may cause huge disparities in allotments among similar customers. This will create serious management problems for the water supplier because many requests for exemptions will be filed, and many people will perceive the system as unfair.

Percent Reduction Allotment: *all account types*

+   easy to determine and administer

+   establish minimum/maximum amounts to limit extremes

+/–   usefulness for nonresidential varies based on efficiency

–   penalizes conservers

–   rewards "above average" users

–   promotes water use during non-shortage periods

The *percentage seasonal allotment* is similar to percentage reduction except that the consumption reduction goal varies depending on the time of year. Both of these methods inadvertently reward past wasteful behavior and penalize past conservation by using previous demand levels in the computation of rationed allotments.

Per connection allotment (residential only) establishes a customer's water consumption goal on a unit basis (such as the number of bedrooms per single-family home or multi-family unit) calculated from an estimate of essential uses. A per connection basis is easier to determine than a per capita basis but may introduce unfair allocations because there is no relationship between historical use, customer characteristics, and how many people live at the residence.

Per Connection Allotment: *residential*

+   easy to establish allotments

–   no relationship between customer characteristics and water use

–   not equitable

–   does not recognize historical use

Per capita allotment (residential only) provides a fixed amount of water per person. This method results in significant work for agency staff, both in determining the number of residents per home and in changing allotments as the number of residents per home changes. It is difficult to equitably provide allocations for anything other than essential inside use.

Per Capita Allotment: *residential*

+ suitable for extreme shortages

+ equitable base allotment, sewer charges on number of residents

− must determine and update per-account occupancy

− water for essential inside use only

− does not recognize historical use

Hybrid per capita/percentage allotment programs have allowed limited outside irrigation, distinguished between single-family and multifamily dwellings with different water use requirements, and still produced 35–45 percent reductions. Customers prefer a fixed allocation within which they can determine their own water use priorities. The hybrid provides water for inside use and a percentage of the five-year average for outside use. Geographic information systems now allow the outside-use portion of the hybrid system to be based on the landscaped area served by each meter, and some agencies use real-time weather data to set the actual monthly outdoor allocation. A maximum per customer allocation is necessary, however, to limit the amount of water allocated to large parcels. Otherwise, the community may not see the allocation method as equitable. The type of rationing selected depends on three factors: (1) the amount of water available for health, safety, and sanitary purposes, commercial-industrial uses, and agriculture and landscape irrigation; (2) the seasonal variation in water consumption (usually a function of irrigation demand); and (3) the degree of homogeneity among consumer types. Where water is in extremely short supply and no water is available for irrigation, the fixed allotment approach usually works best. Where some water is available for landscape irrigation, a plan that permits the customer more water in the dry season, a hybrid per capita basis is preferable.

Hybrid Per Capita/Percentage: *residential*

+ equitable—recognizes variety of uses

+ flexibility—suitable to all stages

+ provides customers greatest control

+ recognizes factors like lot size, historic use, and economics

− additional staff/computer work to determine allotments

− requires more public education

When rationing is in effect, water suppliers may want to make special efforts to help customers save trees in their landscapes. Mature trees take longer to establish than smaller landscape plants, and their value is greater. Trees also provide shading and cooling and help to keep the air cleaner. If the water supplier permits some landscape irrigation, it may advise customers to irrigate trees because they are the most valuable component of the landscape. Even if no landscape irrigation is permitted, customers might be advised to use graywater to keep trees alive. Encouraging customers to install rain barrels or other rain catchment systems can also augment irrigation needs when community water supplies are restricted.

Enforcement

During significant shortages, a call for voluntary conservation may not bring sufficient reduction of water use, especially when water consumption is to be reduced by 15 percent or more. In such cases, mandatory conservation measures (such as restrictions or rationing), which are enforceable under the authority of special ordinances or revised rate schedules, may be necessary. It is recommended that utilities have an escalation policy in place that will clearly define how violations of drought policy will be managed. Table 3-3

Table 3-3   Examples of drought/emergency conservation penalties

Violation Occurrence	Penalty	
	Prohibited use	Excess use
First	Written warning by regular mail and/or door tag.	Written warning by regular mail.
Second	Written warning delivered by utility representative who will offer conservation tips and approved retrofit devices. $25 charge to account on second notice in one-year period. (Note: Some agencies double their fines during declared water shortages.)	Surcharge if allotment is exceeded. (Note: Some agencies double their surcharge rates during declared water shortages.)
Third	Written warning delivered by utility representative who will offer conservation tips and approved retrofit devices. $100 charge to account on third notice in one-year period.	Surcharge if allotment is exceeded.
Fourth	Flow restrictor (1 gpm) installed for 48 hours; installation and removal charges assessed (recommended for severe droughts).	Surcharge if allotment is exceeded.
Additional	$500 charge to account on fourth or subsequent violation within one-year period. Shutoff, plus reconnection charge.	Shutoff, plus reconnection charge.

summarizes penalties that can be used to enforce such programs. The most severe violations may call for shutting off service. Less extreme infractions may be handled with a rate schedule that imposes financial penalties for excess use.

Education, Citations, and Fines

Many water suppliers have noted that the availability of enforcement mechanisms is the most important feature of a WSP and that application of enforcement procedures is rare. Nevertheless, it is important that customers know that those who choose not to cooperate will be dealt with firmly and to give customers an opportunity to report water waste, either through a smart phone app, via email, or by phoning their customer service department. This way, the consumer is assured that the plan is uniformly applied and fair to all. Communicate the supplier's resolve to enforce the regulations at the start of restrictions.

A common enforcement mechanism is to use water-waste patrols, frequently referred to as "water cops" or "waste busters." They usually issue warnings for the first violation. Subsequent violations are subject to fines, and if still uncorrected, installation of a water flow restrictor at the customer service connection. Water cops enforce the water shortage restrictions and water waste rules. The goal of water cops is to use education to help customers save water, not merely to penalize violators.

Restrictions on the days when landscape irrigation is allowed have not always been successful when not enforced. Some residents water on the designated days regardless of whether the landscape needs it. Others over-irrigate their landscapes in the hope the additional water will keep their landscape alive. This overuse cannot be entirely controlled by patrols as the overwatering may be occurring in the customer's fenced backyard.

Landscape irrigation runoff is easily detected by water patrols when it occurs on front lawns and public and business landscapes and patrolling occurs early in the morning or late at night. The value of these patrols is to help customers understand and operate their irrigation systems. The patrols are also a visible reminder to the community of the seriousness of the situation.

Patrols are particularly necessary when there are restrictions on the time of day when landscape irrigation is allowed. The patrols should be scheduled to do most of their patrolling in the evenings and early mornings. Restrictions on midday watering mean that many residents with automatic sprinklers will schedule watering for when they are asleep, and sprinkler malfunctions may go unnoticed. Time- and date-stamped photos or videos made during evening or early morning patrols have been a useful tool to demonstrate to property managers and nonresidential property owners that their irrigation systems need repair or adjustment. Photos also reduce conflict and blame about the nature and cause of water waste cases and usually accelerate resolution of problems. Good software now exists to help water suppliers manage water waste enforcement caseloads, correspondence, and documentation.

Monitoring customers for compliance with mandatory measures that are not strictly consumption-related is complex. Most water suppliers rely heavily on peer pressure and observations by the public and by water-supplier field employees during their regular work schedule. Also, city or county employees whose daily work routine requires them to move about the community can be empowered to issue citations, although these employees are often reluctant to fulfill this role. These types of employees include supervisors of street and wastewater departments and inspectors for building, plumbing, electrical, construction, and health services. This is an effective method of covering the service area at minimum expense and with little interruption of the employee's regular duties. Police are not widely used unless there is a problem with a specific customer.

Most jurisdictions provide an appeal process for customers, and some offer alternatives to fines, including water conservation classes, interior and exterior water use efficiency retrofits, the application of fees to a professional landscape evaluation and water audit, and participation in irrigation upgrade and landscape conversion programs.

Texas Cities Enforce Drought Ordinances

Citations issued by the city of Corpus Christi, Texas, for violations of the water conservation ordinance represent a misdemeanor charge punishable by a fine not to exceed $200. The San Antonio, Texas, water conservation plan provides for a special team of civilian field investigators to enforce the drought response ordinance when the highest-level drought stage is reached. These field investigators will be drawn from the fire, building, health, wastewater, public works, and planning departments. Investigators will be empowered to issue both warning and regular citations to violators. Enforcement powers needed by a water supplier should be clearly described in drought ordinances.

Customer hotlines, phone apps, and online reporting websites have also been useful in identifying repeat violators. They can be integrated with the agency's work order system to send a monitoring official to visit the reported address to issue a warning or citation as necessary. Hotlines and online reporting greatly assist the agency's credibility in enforcement if reported violations are quickly followed up by enforcement. One drawback reported is that some neighborly feuds generate repeated calls to hotlines. The agency staff will have to identify such patterns and stop responding to calls where no violations are observed. Despite this perceived drawback, the reduced time to address violations because of public reporting helps enforcement.

Flow Restrictors

Some customers will continue to exceed their allotment regardless of the amount of their water bill unless the fines are significant. Even one customer that refuses to support the community's efforts to reduce water use can undermine the essential community belief in equity. At some point, the media will contact water suppliers with a request for information on customers with the greatest water use or largest bills. Water suppliers have the legal authority to enforce drought regulations by terminating service. Some suppliers have instead chosen to install flow restrictors on noncooperative customers. Flow restrictors can be manufactured by the supplier to provide, for instance, a one-gallon-per-minute flow—allowing only enough water for health and safety needs. Flow restrictors may not be allowed in jurisdictions where required fire suppression sprinklers are on the same supply line as noncooperative home or business.

Exceptions and Appeals

No WSP can account for all situations. When enforcing water restrictions or rationing, it is important to establish an administrative process for allowing exceptions to the regulations where the need for additional water or special consideration is justified. A similar process must be instituted to handle any cases or appeals by those who seek to challenge the enforcement action.

One creative solution used by the cities of Sacramento and Santa Cruz in the 2014–2017 drought was to implement a "Water School" program. In the same way that drivers who are cited for moving violations must take driver's education courses, Water School is a way to educate the community about where their water comes from, the severity of the drought, and regulations that are in effect and to offer an alternative to paying significant water waste or rationing fines, thereby reducing the caseload of appeals.

Feedback to Customers

To help customers understand how they are doing in meeting their conservation requirements, simple instructions should be provided on how to read meters and to convert billing units to gallons.

A comparison of actual water use with allocation/target should be included on the bills. Also, the customer's allocation for the next billing period should be included on the bill. If the customer has a yearly allocation, information should be provided on the year-to-date use on the bill. To ensure that customers are aware of conservation requirements and whether they are meeting or exceeding them, email notifications should be sent that remind customers to check the change in water use displayed on their bills, and this information should be displayed on online billing websites.

A variety of cell phone apps and other similar tools are now available and can be used to help provide customers with up-to-date information such as how they are doing with respect to meeting their water conservation requirements.

Demand Offset Programs (for New Developments During Drought)

During the early phases of a water shortage, water suppliers are often extremely reluctant to stop providing new connections. To alleviate the public's concern during a water shortage, a number of suppliers have demand offset programs that require the builder to fund water conservation projects at a level that is one, two, or three times the demand of the projected project. A number of agencies have demand offset programs (AWE 2015).

A one-to-one offset puts the existing community at a disadvantage. Although the developer has offset the new demand, this has been accomplished by using some of the community's extra water. When the next drought occurs, there will be less extra water and the new development, which is already water efficient, will increase demand.

This increased demand can be compensated for by having a greater than one-to-one offset. The developer would fund conservation of more water than the amount the new project would use. This would mean that the new project would make the community more resilient in the face of a water shortage. Cities as diverse as Santa Monica and San Luis Obispo, California, and San Antonio, Texas, have required offsets as part of their ongoing efforts to reduce demands.

Irrespective of the approach used to manage new construction, it is important to recognize that if a water shortage persists, the additional demand resulting from new connections approved during the shortage could result in economic harm to existing customers.

REFERENCES

Australian Treasury. 2004. The impact of the 2002 drought on the economy and agricultural employment – Treasury.gov.au.

AWE (Alliance for Water Efficiency). *Water Offset Policies for Water-Neural Community Growth: A Literature Review & Case Study Compilation.* www.allianceforwaterefficiency.org/water-offset-report-Jan-2015.aspx (accessed Oct. 24, 2018).

AWWA (American Water Works Association). 2016. M36 *Water Audits and Loss Control Programs, Fourth Edition.* Denver, Colo.: AWWA.

Heberger, M. 2012. "Australia's Millennium Drought: Impacts and Responses." *The World's Water,* 97–125.

Mini, C.; Hogue, T.S.; and Pincetl, S. 2015. "The Effectiveness of Water Conservation Measures on Summer Residential Water Use in Los Angeles, California." *Resources, Conservation, and Recycling,* 94:136–145.

NYC DEP (New York City Department of Environmental Protection). 2019. "Operations Support Tool." https://www1.nyc.gov/html/dep/html/drinking_water/forecasting_reservoir_levels_ost.shtml (accessed Feb. 20, 2019).

This page intentionally blank.

Step **4**

Establish Triggering Levels

No matter how seemingly reliable a water supply is, drought triggers are needed to protect against potential shortfalls due to extended periods of low water supplies. Much of the water industry relies on the concept of safe yield for planning. Experience in recent years in various parts of the world has shown that any particular drought might last longer than the one used to determine the safe yield. For river supplies that lack reservoir storage, early warning triggers are especially critical to guide use restrictions and activation of emergency backup supplies since drought onset may be quick and result in acute impacts. For groundwater supplies, drought may take longer to develop and lead to more subtle impacts that require unique drought triggers and metrics.

This section provides an overview of drought triggers that are needed to balance supply and demand during water shortage periods. These triggers can vary from simple in nature (e.g., days of supply remaining) to medium in complexity (e.g., seasonal rule curves) to sophisticated (e.g., probability-based using inflow forecasts). Through the growing need to better quantify risk as water suppliers are dealing with increased demands on the water supply and a changing climate, communities are increasingly transitioning from simple to sophisticated drought triggers. Examples from several communities are showcased to provide some details about how such plans have been developed and how the triggers are implemented. This section also addresses lessons learned from when triggers were revised due to less than optimal results. Finally, considerations for communicating the triggers and the rationale underlying them will be provided.

TRIGGER MECHANISMS

Optimal drought triggers will effectively balance system reliability and the likelihood of continuing drought against the costs and challenges associated with new water supplies versus imposing water use restrictions and other drought emergency actions. Some considerations when developing the triggers include the following:

- What is the likelihood of the reservoir elevation, stream flow, or groundwater elevation dropping below a certain level?
- Are there additional supplies available under contract, and what would be the cost of increasing them?
- How often are water use restrictions needed?
- How large should the restrictions be and how long must they stay in effect?
- What is the appropriate timing for imposing restrictions (voluntary, mandatory, surcharge)?
- When should backup supply sources be activated?

For a water shortage plan to be successful, it must consist of suitable triggers that

- identify all droughts and provide enough lead time for the corrective action(s) to be effective—running out of water is not an option;
- are activated no more frequently than deemed acceptable by the water supplier and the community;
- minimize false alerts or times when supplemental water supplies are purchased or water use restrictions are unnecessarily imposed—false alerts can be costly to a water supplier and its customers and can damage the supplier's credibility with the public; and
- are based upon resources available to the supplier and weather criteria—these should be adopted ahead of time by a policy body, and then a duly authorized manager implements the water shortage plan (WSP) as they are reached.

Because it is unreasonable to expect that every possible combination of factors influencing system performance during drought can be predicted, triggers should be flexible and mitigation actions robust to ensure a "no regrets" set of solutions. A WSP tailored to the historical drought of record may perform well for droughts of similar magnitude and duration, but they may leave a water supply vulnerable to more severe droughts or those that develop with different spatial and temporal characteristics. A dynamic plan that allows for managers and operators to reevaluate conditions over time and modify responses as conditions warrant will provide the most flexibility and can help a water supplier overcome the considerable uncertainty in future conditions that can complicate drought mitigation planning.

Whatever parameters are used for trigger mechanisms, they should be ones that can be assessed frequently. The analysis of such information should be readily available to decision makers in a timely manner. Using such a quantified system, advancing through drought stages can be almost automatic. When complex trigger mechanisms are used, however, some uncertainty may arise as to whether to initiate a given drought stage. For example, reservoir levels are low, but a water transfer is being negotiated. Assign the resolution of such "gray area" decisions to a specific individual or group that is clearly responsible for making these difficult decisions. *Early implementation of demand-reduction measures should be the guiding rule—not "hoping for rain."* Use of simulation models as described in this chapter can be used to see how the triggers would have worked in past droughts, reducing those potential uncertainties.

Certain triggers may work well only over a limited range of system demands. If demands increase to the point that the triggers are invoked too often or system storage is not preserved in the worst droughts, the triggers will need to be updated, potentially in combination with increases in system capacity or more serious demand-reduction measures.

Table 4-1   Triggers by type of supply source

Water Supply Source	Key Indicators	Candidate Trigger
Groundwater	Groundwater levels, historic and recent production levels	Groundwater stage (elevation)
River	Precipitation, snow pack, soil moisture in watershed	River stage (elevation)
Surface water	Storage levels, precipitation, snow pack, soil moisture in watershed	Reservoir stage (elevation)

Table 4-1 lists various types of sources of supply and the key indicators and triggers that may be applied. Additional examples are also provided in this section. Proactive measures and ordinances can be established with tiered restrictions based upon triggering mechanisms such as aquifer monitoring well levels, river flows, or surface water levels.

Groundwater

Groundwater supplies typically recharge at slower rates than surface water, and drought triggers are usually based on the drought of record. Monitoring well levels is a common approach used to determine triggers, although site-specific conditions can drive water depths lower in particular aquifers or zones of heavy pumping. Rules are managed at state, regional, or district levels depending on the aquifer, and an individual water supplier may be given triggers based on legislative or administrative decisions.

Where a water supplier is managing resources absent of such external regulation, the following considerations are paramount:

- historical data on aquifer levels and related to previous droughts;
- water quality changes as depth to water increases;
- cost of pumping or lowering pumps when water levels decline;
- impacts of expanding cones of depression and effects on other wells; and
- costs and impacts of additional interconnect or conjunctive uses if such are possible.

Some examples of groundwater regulations for drought periods include water districts in the Ogallala Aquifer region of the high plains states in the United States (USDA n.d.); the Edwards Aquifer Authority in central Texas (2019); adjudicated basins in California (California Department of Water Resources 2019); and the Groundwater Management areas of Arizona (Arizona Department of Water Resources 2018).

River Supplies

River surface water supplies generally have less flexibility than reservoir or groundwater supplies. Riparian rights for all users on a river system mean that sufficient water needs to be left for downstream flow. In some states, environmental flows are also protected. In addition, when hydroelectric projects are located in a river system, these supplies are often subject to instream or pass-by flow restrictions, limiting withdrawals under low-flow conditions. Accordingly, triggers are often used to implement drought mitigation actions as early as possible. For systems with multiple supply sources, run-of-river withdrawals may be minimized during low-flow conditions in favor of stored water. Low-flow triggers may be used to implement conservation measures. For example, a staged WSP may be activated based on a running average river flow compared to a pre-selected threshold. This threshold could be expressed as a gage height, a flow rate, or a percentage of an allowed withdrawal. Triggered drought responses may include voluntary use restrictions under warning stage and mandatory restrictions under emergency stage. Since all users within a watershed are likely facing similar challenges, a regional, staged approach is advisable. Regional bodies can coordinate on more than just drought conditions. Basic conservation measures applied under normal conditions, consideration of total regional withdrawals, and flow-triggered emergency actions can help water suppliers optimize use of available water while protecting instream uses.

Surface Water Storage

Simple triggers, such as days of supply remaining (where inflow is ignored), can be calculated without use of simulation models and are easily understood by the public as the amount of system storage divided by the demand. However, this can be very misleading and might provide a false sense of security (or risk) since inflow is ignored. With recent improvements in the National Weather Service's medium-to-long-range (2–40 weeks) hydrologic ensemble forecast products (e.g., Advanced Hydrologic Prediction Service, or AHPS; Hydrologic Ensemble Forecasting System, or HEFS), it is possible to calculate days

New York City Delaware River Basin Storage, Sept. 4, 2018

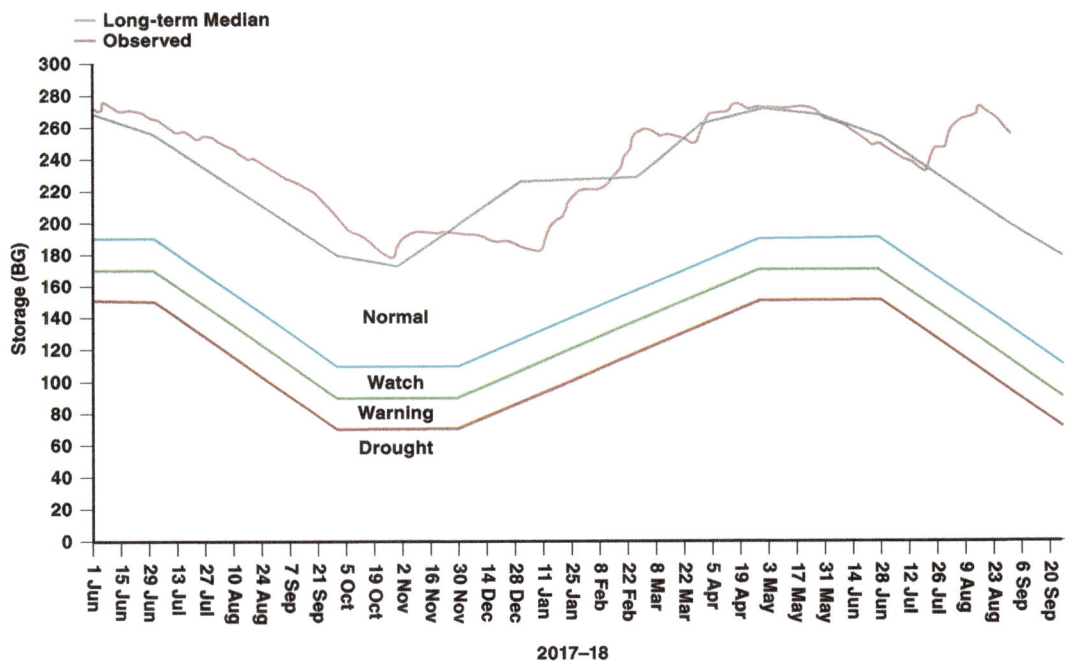

Useable	Cannonsville	Pepacton	Neversink	Total
BG	87.9	133.2	34.1	255.2
%	94.1%	95.6%	98.4%	95.4%

BG above drought watch =	122
BG above drought warning =	142
BG above drought =	162
BG above daily storage median =	57
BG above one year ago =	27

Storage data are provisional and provided by the New York City Department of Environmental Protection, Bureau of Water Supply. The period of record represented by the long-term median is June 1967 to May 2013. Drought Watch, Warning and Drought are defined by Figure 1 of Article 2 in the Delaware River Basin Water Code 18 CFR Part 410.
As of June 1, 2018, the NYC Delaware reservoir statistics have been changed to reflect the 2016 USGS bathymetry tables.

Adapted from Delaware River Basin Commission 2019. Note: For the most recent storage level, visit http://www.state.nj.us/drbc/hydrological/reservoirs/nycdrb-reservoirs.html

Figure 4-1   Delaware River Basin storage levels

of supply remaining that includes a forecast of surface water inflows, based on an assumed level of risk.

Seasonal rule curves that are tied to system storage are typically developed using a simulation model, and curves that would have worked in past droughts (usually based on simulation of historical hydrology) provide greater confidence they will work equally well in future droughts. These curves are also easy to interpret. An example for New York City is shown in Figure 4-1 based on its three-reservoir supply in the Delaware River Basin.

However, given that all droughts are different—in timing, duration, and intensity—explicit accounting of risk using probability-based drought triggers is the most effective means of detecting and responding to drought. In this case, the trigger would be based on the probability of the system storage reaching a certain level in the future. This trigger would factor in the starting storage and inflow conditions at the time the forecast is made, thus providing the most accurate assessment of all the triggers in terms of drought risk. Such triggers would need to be developed and implemented using simulation models. Communicating such triggers to the public is more challenging but made easier by growing use and awareness of probabilities as it relates to short-term weather forecasts.

Following is an example of how a storage forecast and the relevant trigger criteria—x percent chance of hitting y percent storage in z weeks—would be used to determine if the trigger is met: if the trigger is based on a 20 percent chance of being below 50 percent storage in 12 weeks and the plot shows less than a 10 percent chance of that occurring, the trigger has not been invoked. When drought triggers have been determined, a mathematical model can be used to predict water supply conditions during a drought in real time (see Figure 4-2). In the following weeks, the operator would run another forecast as storage level declines and the likelihood of hitting the trigger increases.

Reservoir storage 12 weeks in the future

When drought triggers have been determined, a mathematical model can be used to predict water supply conditions during a drought in real time.

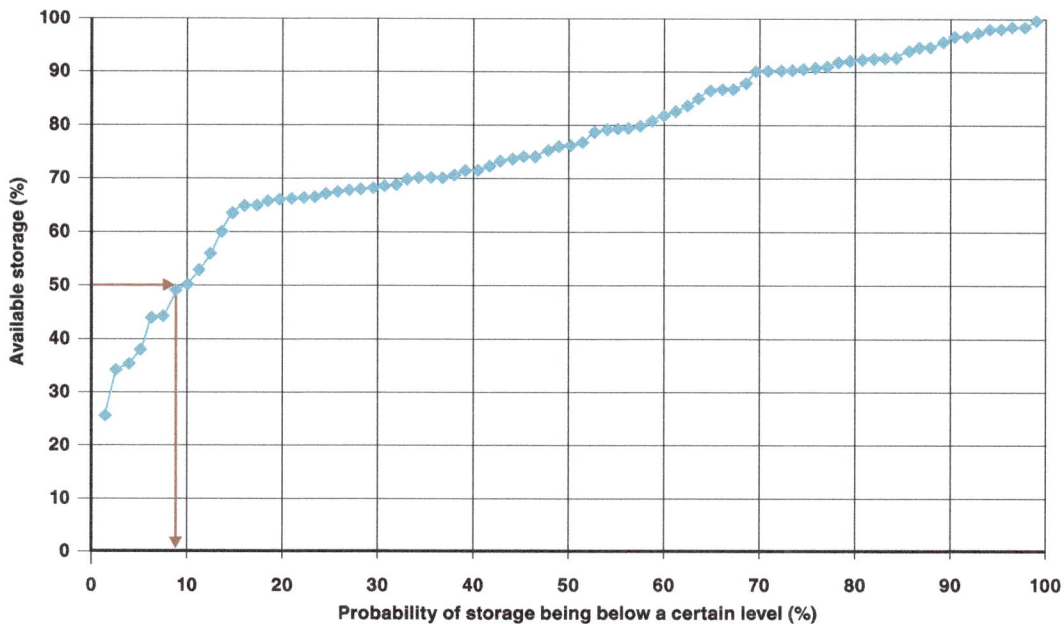

Source: AWWA

Figure 4-2   Model for determining probability of drought

Alternatively, say there is a 20 percent chance of the reservoir elevation dropping to 2,575 ft 10 weeks in the future. In this case, the trigger is activated and corrective action such as demand reductions will be imposed. Rather than waiting until the reservoir reaches that level, the trigger is designed to act on the probability of reaching that level in some specified timeframe. Thus, operators are being proactive in the drought response. Similarly, if reservoir inflows are likely to increase as part of a typical refill period, even if reservoir storage is lower than usual, the forecasts may tell operators to avoid taking action since the likelihood of refill is sufficiently high. The City of Asheville, N.C., is an example of a utility using this forecast-based approach to trigger implementation (Dundas 2016).

Combined Sources of Supply

Many water suppliers have more than one source of supply, and actual management options may include switching reliance from one to another based on circumstances. Many water suppliers in California depend on riparian or reservoir storage when the water year is adequate. When drought conditions curtail those supplies, they shift to groundwater pumping, if available, or supplemental supplies they have under contract. In other circumstance, such as in Waco, Tex., where treatment of groundwater historically has been less expensive than surface water treatment, the order is reversed, with more surface water being used when groundwater supplies are stressed.

Figure 4-3 shows how the status of several different water supplies can be used in setting triggers and determining when to invoke drought stages. It illustrates an actual situation experienced by Greater Vernon Water, a supplier of the Regional District of North Okanagan (RDNO), British Columbia (Clarke Geoscience Ltd. 2011). In April 2011,

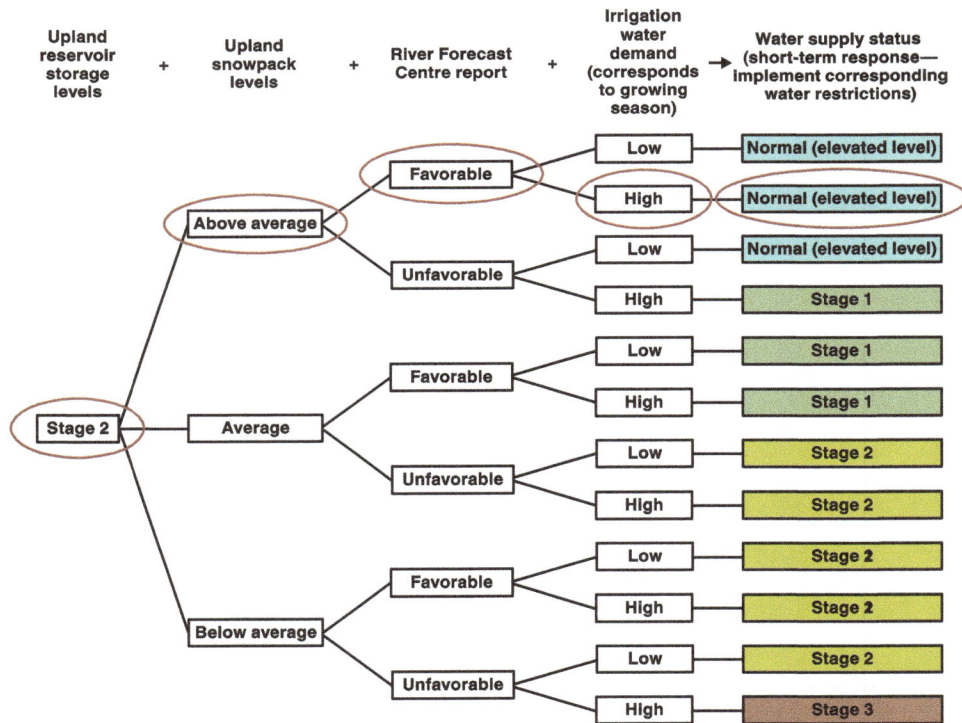

Adapted with permission from Greater Vernon Water Utility Drought Management Plan (Update revised October 2011) prepared by Clarke Geoscience Ltd., Kelowna, B.C.

Figure 4-3   Multiple supplies affect drought stages

reservoir supply levels were at Stage 2 under the drought management plan (DMP). However, this level was not the only consideration in determining the restriction stage for the supplier; a decision tree was used to determine what overall restriction stage was needed. As snowpack was above average and the weather prediction was for a wet summer—a favorable forecast—the supplier determined that moving to Stage 2 restrictions was unnecessary. Staff watched the restriction trigger parameters carefully as they moved into the summer higher water demand season, as high demand coupled with a sudden dry period could push the water supplier toward higher restrictions. As the year progressed, they were able to stay at "normal" restrictions, albeit with an elevated level of awareness that conditions could change.

The DMP includes operational considerations at each stage. At normal (elevated) and Stage 1, utility staff members

- increase efforts to communicate water conservation messages to the public and encourage voluntary conservation and
- look at ways to adjust reservoir and distribution operations, such as by using alternative sources that are less affected by drought, to bring reservoir storage back into the normal range.

When the DMP was revised in 2010–2011, a stage was added to allow for earlier warning that drought conditions were occurring and also to guide staff in undertaking actions that would improve reservoir storage conditions while minimizing impacts on customers.

Causes of Delayed Implementation of a Stage

Water suppliers will face enormous pressure to not declare a water shortage. Imposing restrictions, significantly increasing water rates, or rationing often results in upset customers that rightly or wrongly anticipate damage to their businesses, homes, and lifestyles. Political leaders need clearly defined triggers to make decisions when there is a water supply problem. Plans that depend on a political body to determine when a drought has begun often fail. They may delay authorizing restrictions due to perceptions and pressure from local businesses, thus resulting in greater drawdown of water resources and the need for more restrictions as a drought deepens.

At the same time, there will be pressure to act during a drought to address environmental impacts on water resources, including possibly endangered or threatened species. Low reservoir levels and low river flows also reduce or prevent recreation and impact agriculture and can create aesthetic problems. All of these situations are exacerbated by delaying the implementation of a new stage of the WSP with its cutbacks on demand.

It is important, then, that triggers be clearly defined and documented as part of the adopted WSP.

Include Flexibility

Some WSPs have included rigid requirements for moving from one stage to another, including board or city council approval. This has resulted in rapidly worsening conditions when drought response was delayed or operational measures were insufficient to slow the depletion of reserves. It is better to give the water supplier some flexibility and authority to adjust the trigger and action levels based on some predefined guidance. Correlate the number of successive levels of drought stages with a series of realistic deficit-reduction goals. Most communities use three to five stages. Fewer than three stages would likely require dramatic changes between the first and second stages. More than five stages may incur frequent transitions that could reduce the effectiveness of deficit-reduction measures as they are introduced.

Additional indicators of drought can be incorporated into a water shortage plan, not the least of which are precipitation, streamflow, groundwater levels, Palmer Drought Severity, and local Drought Monitor (see Step 2 of this manual). A weighted combination may be derived from a combination of all of the relevant factors, including current system storage and ensemble streamflow forecasts.

When the Drought Recedes

As a drought recedes and water supply conditions improve, WSPs need to have an exit strategy for the various stages. In Seattle, as soon as actual and forecast supply conditions substantially improve, Seattle Public Utilities (SPU) will either inform the public of the return to normal use of water or will inform them that SPU is moving from one stage to a lesser stage of the plan. This latter process would occur until there is a return to normal operations. Stages can be skipped in this process as conditions and forecasts warrant. See Step 7 of this manual for a more detailed description of the steps to take at the end of a drought.

REFERENCES

Arizona Department of Water Resources. Arizona Drought Management-Overview. https://new.azwater.gov/drought (accessed Nov. 5, 2018).

California Department of Water Resources. 2019. Adjudicated Areas. https://water.ca.gov/Programs/Groundwater-Management/SGMA-Groundwater-Management/Adjudicated-Areas (accessed March 20, 2019).

Clarke Geoscience Ltd. 2011. Greater Vernon Water Utility (GVWU) Drought Management Plan. www.rdno.ca/docs/drought_management_plan.pdf (accessed March 7, 2019).

Delaware River Basin Commission. 2019. NYC DRB Reservoirs. http://www.state.nj.us/drbc/hydrological/reservoirs/nycdrb-reservoirs.html (accessed March 7, 2019).

Dundas, J. 2016. "Predictive Model Helps Asheville Manage Water Supply During Drought." http://coablog.ashevillenc.gov/2016/11/predictive-model-helps-asheville-manage-water-supply-during-drought/ (accessed March 7, 2019).

Edwards Aquifer Authority. 2019. "Critical Period/Drought Management." https://www.edwardsaquifer.org/groundwater-users/groundwater-permit-holder/critical-period-drought-management (accessed Nov. 5, 2018).

USDA (US Department of Agriculture). n.d. Natural Resources Conservation Service – South Dakota. "Ogallala Aquifer Initiative." https://www.nrcs.usda.gov/wps/portal/nrcs/detailfull/sd/home/?cid=stelprdb1048809 (accessed March 20, 2019).

Step **5**

Develop a Staged Demand-Reduction Program

The demand-reduction measures that will be used in each stage to produce the necessary water savings should be identified and detailed. Supply augmentation was discussed in Step 4, when the water shortage stage triggers were developed. In Step 5, demand-reduction measures, including those described in Step 3, are correlated with customer water use characteristics and the projected savings are quantified. Program design is evaluated for effectiveness, timeliness, and cost. Groups of water-saving measures will be associated with progressive levels of supply shortage.

CRITERIA FOR DEMAND REDUCTION DURING A WATER SHORTAGE

There are several criteria for deciding which demand-reduction measures are appropriate to reduce demand during a water shortage.

- Timing: Can the measures or actions produce results in the necessary timeframe?
- Magnitude of savings: Will the measures or actions result in enough water savings to make a meaningful difference? For example, will it reduce demand to the level the impaired water system can support?
- Season: Are the actions or measures relevant to the time of year? For example, banning or severely reducing lawn watering during the summer versus during winter.
- Costs: How severe are the cost implications of the measures to the customer, including local business and industry, relative to the need for action? Note: While

there could be costs to certain customers, particular actions may still be necessary for public health and safety.

ESTABLISH STAGES

The most common approach to managing water demand during a water shortage is to use a staged approach, with increasing levels of savings in each successive stage. Each new stage of reductions will begin when a trigger is reached, as determined in Step 4. An example of staged reduction goals is shown in Table 5-1, and an illustrative example program is described in Tables 5-2a and 5-2b, which present the customer reduction goals established by the City of Santa Cruz (2015), California, in its 2015 Urban Water Management Plan. For Santa Cruz, the customer reduction goals for all but the first stage were developed by evaluating the composition of demand for each major group and dividing it into three basic usage priorities. These priorities are, from highest to lowest, (1) health and safety, i.e., domestic and sanitary uses; (2) business and industrial uses; and (3) irrigation and other outdoor uses. Normal demands were then scaled back in accordance with the schedule described in this section to meet the overall reduction needed. At each level of shortfall, public health and safety usage is always afforded the highest priority by cutting back on it the least and by cutting back irrigation and outdoor uses the most.

Stage 1 relies primarily on voluntary action by customers, tied to an outreach campaign. These actions are taken in anticipation of the drought continuing and the community benefiting in some cases from increased carryover storage. Pumps can continue to operate, and more water is available for essential uses. Subsequent stages are in response to increasing supply shortages. Stage 2 typically uses some mandatory measures, and Stages 3 and 4 typically involve water rationing. Stage 4 includes extensive restrictions on water use and would be initiated only in extreme circumstances. Efforts made during the first three stages to avert reaching Stage 4 will save the customers and the water suppliers from the potential hardships of extreme shortages.

The estimated percent demand reduction in Table 5-1 for each drought stage was derived from savings achieved in previous drought situations by comparable water shortage plans. The typical demand-reduction goals for staged plans normally range from 10 percent in the first stage to as much as 50 percent in the last stage.

When setting demand-reduction targets, it is important to do an evaluation to determine whether targets can be achieved without compromising health and safety. Different targets may be appropriate for different customer classes. Customer use data for at least the past five years should be reviewed before deciding on reduction targets for each customer class. For example, a utility may decide that it wants to target the single-family and multifamily class for a 50 percent demand reduction under Stage 4. But after carrying out an analysis of the five-year average use for each class, the utility finds that a 50 percent reduction in multifamily use would reduce consumption to levels below its desired health

Table 5-1 Example stages with demand-reduction goals

Stage	Water Shortage	Demand-reduction Goal, %
1	Minimum	10
2	Moderate	20
3	Severe	30
4	Critical	40

Table 5-2a   Santa Cruz customer reduction goals by stage

Customer Class	Normal Demand (million gallons)	Stage 2 % Reduction / Amount delivered	Stage 3 % Reduction / Amount delivered	Stage 4 % Reduction / Amount delivered	Stage 5 % Reduction / Amount delivered
Single Family Residential	1,031	16%	27%	38%	52%
		864	753	639	495
Multiple Residential	524	13%	22%	31%	45%
		454	411	361	287
Business	438	5%	8%	13%	30%
		416	402	381	307
UC Santa Cruz	132	15%	24%	34%	48%
		113	100	87	68
Industrial/ Agriculture	82	5%	10%	15%	33%
		78	74	70	55
Municipal	48	24%	43%	59%	72%
		36	27	20	14
Irrigation	110	36%	66%	88%	100%
		70	37	13	0
Golf Course Irrigation	106	27%	49%	66%	80%
		78	54	36	21
		Overall reduction in each stage:			
		15%	25%	35%	50%
Total	**2,473**	**2,111**	**1,861**	**1,607**	**1,247**

Table 5-2b   Santa Cruz customer reduction priorities by stage

	Reduction in Water Delivery by Usage Priority (percent of normal distribution)			
Stage	Magnitude of Water Shortage	Health/Safety	Business/Industry	Irrigation/Other Outdoor Uses
2	15%	95	95	64
3	25%	95	90	34
4	35%	90	85	12
5	50%	75	67	0

and safety allocation of 35 gallons per capita per day (gpcd). Based on this analysis, the utility instead sets its multifamily target to 40 percent so that health and safety for residents in the multifamily class can be maintained.

MEASURES

The impacts of implementing short-term versus long-term demand-reduction measures should be considered. Short-term measures may be cheaper and faster to implement, but long-term measures may provide permanent increased water supply reliability; for example, providing shower timers (short-term) versus replacing inefficient toilets with efficient ones (long-term).

Demand-reduction measures seek to reduce water use through plumbing fixture replacement, fixture leak detection and repair, water audits to reveal more efficient or alternative ways of using water, improved landscape irrigation practices, and use restrictions specific to the customer class. Table 5-3 provides examples of demand-reduction measures by customer category.

Table 5-3   Demand-reduction measures

Customer Category	Examples

Existing Residential	1. Interior
	  a. Public information
	  b. Residential water audit
	  c. Fixture leak detection and repair
	  d. Plumbing fixture replacement
	  e. Appliance replacement
	2. Exterior
	  a. Public information
	  b. Landscape water audit
	  c. Turf irrigation guidelines/irrigation timer settings
	  d. Pool covers and refill restrictions, hose nozzles
	  e. Landscape irrigation improvements and repair
	  f. Use restrictions (day or time of use)
	  g. Graywater use guidelines

Existing Commercial and Institutional	1. Employee information programs 2. Interior water-use audits 3. Landscape irrigation audits 4. Plumbing fixture repair and replacement 5. Irrigation system repair and improvement 6. Specific use restrictions
Existing Industrial	1. Employee information programs 2. Interior water-use audits 3. Landscape irrigation audits 4. Plumbing fixture repair and replacement 5. Irrigation system repair and improvement 6. Process water audits 7. Process system repair and improvement 8. Specific use restrictions
New Connections	1. Information program 2. Plumbing code changes 3. Restrictions on new landscaping 4. Pool-filling restrictions 5. No new landscaping 6. Connection moratorium

Existing toilet and washing-machine rebate programs can be expanded. While replacement is normally part of a long-term conservation program, it can be implemented quickly if enough financial and human resources are allocated. A water supplier could decide to increase the rebate amount or may decide to offer a direct installation program to all or targeted sectors of their service area.

A 2016 study of residential water use found that the average indoor use was 59 gallons per capita per day (gpcd; WRF 2016). Similar studies of homes with a full range of water-efficient products and features estimated an indoor use of 37 gpcd, 22 gpcd less than average homes. A replacement campaign to reduce consumption in less efficient homes by providing efficient showerheads and dye tablets to identify leaky toilets and by offering rebates on toilets, washing machines, and low-volume irrigation equipment as well as other types of rebates could effect significant water savings.

Some water suppliers offer incentives to replace grass as part of their ongoing demand-reduction programs. This can be expanded and focused on the installation of appropriate landscaping. Other possibilities include promotions of efficient irrigation technology such as drip and micro-spray or high-efficiency sprinkler nozzles. Offering irrigation system audits can ensure that customers are following the utility's water restrictions and can drive them to participate in rebate programs.

During moderate supply shortages, demand-reduction measures can be incorporated as part of the requirements for new connections to the water system. Alternatively, new customers can be actively encouraged to voluntarily adopt demand-reduction measures with a connection fee discount. During severe and critical supply shortages, deferment of new development or at least deferment of landscape installations in new development is justifiable and likely to be something water suppliers will be asked to consider by its existing customers.

Because already permitted construction projects will continue for many months, the short-term economic impacts on the construction trades could be minimal. One innovative solution the Tuolumne Utilities District (TUD) implemented in order to save water on already permitted construction projects was to provide construction crews with TUD's main flushing schedule. The crews would bring water tanks and use the flushed water for construction.

If a supplier does not stop issuing new meters during rationing, a way to reduce the impact of new connections is to enact a *demand-offset program*. Under this program, developers wanting approval for new construction are required to demonstrate that they will conserve, within the community, two to three times the quantity of water the new project will use. (See p. 50 for a discussion of why offsets greater than 1-to-1 are required.) Developers have the option to carry out the conservation themselves or they may be able to contribute a specified amount into the water supplier's conservation fund. These funds can then be used to finance conservation improvements in public facilities, low-income housing, or expand customer rebate programs. This has the double benefit of conserving water and providing assistance to low-income residents or the whole community.

MANAGING CUSTOMER EXPECTATIONS

Mandatory reduction measures are more severe than voluntary measures and generally produce greater water savings. The principal drawback to mandatory measures is customer resentment if the measures are not seen as equitable. Therefore, the mandatory measures should be designed well and accompanied with a good public relations campaign. Customers should understand that their sacrifices are warranted. The communications

The combined level of Lakes Travis and Buchanan determine Austin's conservation stages.

*Stage 4 - As determined by City Manager, system outage, equipment failure, contamination of water source or other emergencies

Adapted with permission from City of Austin

Figure 5-1   Austin Water's drought triggers illustrated

plan described in Step 7 should include messaging to ensure that the water supplier can demonstrate that it is achieving a balance between demand and available supply. Figure 5-1 shows how the City of Austin, Tex., graphically represents their stages tied to the lake level triggers. Use of easy-to-understand artwork can help achieve the "picture worth a thousand words" effect.

Evaluate Demand-reduction Measures

Once water shortage demand-reduction measures are identified, certain information should be generated to provide decision makers with a basis to review and select measures that will be used in the plan. Three methods of demand reduction exist that are generally imposed to affect all customer classes—restrictions on water use practices, price restructuring, and rationing (limits on customer water use).

Basic considerations include

- water savings,
- lead time required to activate and observe results from measure,
- direct and indirect costs, and
- legal or procedural requirements for implementation.

This step of the water shortage planning process is finalized by eliminating measures that are infeasible and arranging the remaining feasible measures in logical groups. Identify the specific application of a water shortage reduction measure, such as mandatory versus voluntary, residential versus nonresidential, etc. Table 5-4 illustrates how the City of Santa Cruz, Calif., integrates its communication and department activities with its expectations of customer demand reductions. The process is fluid in practice, with communication and demand measures modified to achieve the desired level of reduction.

Table 5-4   City of Santa Cruz summary of demand-reduction actions and measures

Water Shortage Condition	Key Water Department Communication and Operating Actions	Customer Demand-Reduction Measures
Stage 1: Water Shortage Alert (0–5%)	• Initiate public information and advertising campaign • Publicize suggestions and requirements to reduce water use • Adopt water shortage ordinance prohibiting nonessential uses • Step up enforcement of water waste • Coordinate conservation actions with other city departments, green industry	• Voluntary water conservation requested of all customers • Adhere to water waste ordinance • Landscape irrigation restricted to early morning and evening • Nonessential water uses banned • Shutoff nozzles on all hoses used for any purpose • Encourage conversion to drip, low-volume irrigation
Stage 2: Water Shortage Warning (5%–15%)	• Intensify public information campaign • Send direct notices to all customers • Establish conservation hotline • Conduct workshops on large landscape requirements • Optimize existing water sources; intensify system leak detection and repair; suspend flushing • Increase water waste patrol • Convene and staff appeals board	• Continue all Stage 1 measures • Landscape irrigation restricted to designated watering days and times • Require large landscapes to adhere to water budgets • Prohibit exterior washing of structures • Require large users to audit premises and repair leaks • Encourage regular household meter reading and leak detection
Stage 3: Water Shortage Emergency (15%–25%)	• Expand, intensify public information campaign • Provide regular media briefings; publish weekly consumption reports • Modify utility billing system and bill format to accommodate residential rationing; add penalty rates • Convert outside-city customers to monthly billing • Hire additional temporary staff in customer service, conservation, and water distribution • Give advance notice of possible moratorium on new connections if shortage continues	• Institute water rationing for residential customers • Reduce water budgets for large landscapes • Require all commercial customers to prominently display "save water" signage and develop conservation plans • Maintain restrictions on exterior washing • Continue to promote regular household meter reading and leak detection

continued

Table 5-4   City of Santa Cruz summary of demand-reduction actions and measures (continued)

Water Shortage Condition	Key Water Department Communication and Operating Actions	Customer Demand-Reduction Measures
Stage 4: Severe Water Shortage Emergency (25%–35%)	• Contract with advertising agency to carry out major publicity campaign • Continue to provide regular media briefings • Open centralized drought information center • Promote graywater use to save landscaping • Scale up appeals staff and frequency of hearings • Expand water waste enforcement to 24/7 • Develop strategy to mitigate revenue losses and plan for continuing/escalating shortage	• Reduce residential water allocations • Institute water rationing for commercial customers • Minimal water budgets for large landscape customers • Prohibit turf irrigation, installation in new development • Prohibit onsite vehicle washing • Rescind hydrant and bulk water permits
Stage 5: Critical Water Shortage Emergency (35%–50%)	• Continue all previous actions • Implement crisis communications plan and campaign • Activate emergency notification lists • Coordinate with California Department of Public Health regarding water quality, public health issues and with law enforcement and other emergency response agencies to address enforcement challenges • Continue water waste enforcement 24/7	• Further reduce residential water allocations • Reduce commercial water allocations • Prohibit outdoor irrigation • No water for recreational purposes; close pools • Continue all measures initiated in prior stages as appropriate

Source: City of Santa Cruz

Evaluate Water Saved by Staged Reductions

The water saved by one of the four stages listed in Table 5-1 will vary by month. Many measures included in the various stages emphasize reducing outside water use. Therefore, their effectiveness will be greater in the warmer months. The public seems to respond with increased efforts to reduce use when rains fail to materialize. Some water suppliers find that a rationing plan expected to save 25 percent of the total demand on an annual average basis actually saves 35 percent or more in the summer; a correspondingly lower rate of savings, perhaps less than 10 percent, occurs during the winter. Other water suppliers may achieve a more uniform savings throughout the year.

Exactly how much water savings can be achieved in any given month is difficult to predict. A service area where most of the water use is residential with a large proportion used for landscape irrigation may have high summer savings relative to the annual average, whereas a service area with low summer irrigation demands may experience much less variation from the predicted annual average savings.

One way to account for the variation in percent savings is to assume that the savings can be scaled to the normal year demand curve. The predicted percent savings for a given stage in the month of interest is based on the ratio of monthly water use to annual average water use. For example, in Table 5-5, the percent savings from Stage 1, expected to average

Table 5-5   Variation in staged reduction savings

Month	Projected Drought Year Demand, acre-ft	Ratio of Monthly Demand to Average Monthly Demand	Savings by Stage, %		
			1	2	4
January	11,570	0.88	8.83	22.06	44.13
February	11,980	0.91	9.14	22.85	45.69
March	12,680	0.97	9.67	24.18	48.36
April	12,930	0.99	9.86	24.66	49.31
May	13,420	1.02	10.24	25.59	51.18
June	16,310	1.24	12.44	31.10	62.20
July	17,370	1.32	13.25	33.12	66.25
August	15,950	1.22	12.17	30.42	60.83
September	12,120	0.92	9.24	23.11	46.22
October	11,980	0.91	9.14	22.85	45.69
November	10,920	0.83	8.33	20.82	41.65
December	10,090	0.77	7.70	19.24	38.48
		Annual savings goal	10.00	25.00	50.00

10 percent, may vary from 7.7 percent in December to 13.25 percent in July. In 8 out of 12 months, the savings may be below the target, but the water savings in June, July, and August may result in achieving the annual reduction target. While people over-irrigate more in the spring and fall, their overall water use is higher in the summer, thus the opportunity for greater water savings. Seasonal differences are not as marked in coastal areas as they are inland. This example can be used as a general guideline but will not be accurate in every service area. Implementation of demand-reduction programs in unmetered service areas presents special problems.

An effective way to reduce landscape water use is to limit the time of irrigation to early morning and evening. Twice weekly irrigation of lawns is generally adequate for most varieties. Having all irrigation occur, for instance, on weekdays will make enforcement easier. Allowing irrigation three or four times a week, as in odd–even days of the week or every other day, may result in little or no landscape water savings. One way to temper the potential impact on peak water demand is to assign watering days based on odd–even numbered addresses.

As demonstrated in Table 5-5, the ratio of monthly demand to average monthly demand will vary based primarily on the seasonal variation in demand for the landscape. The annual water savings goal for each stage is set based on water supply calculations done in Steps 2, 3, and 4. The monthly reduction targets reflect the proportion of overall water demand to be reduced. This shows that the annual target is achieved by greater reductions in the summer months from reduced outdoor water use, averaged with lower savings in winter, when irrigation is not as prevalent.

Lag Time Issues

Water consumption in the 1996 drought in San Antonio, Tex., shows evidence (Figure 5-2) of the lags between announcement of a drought surcharge (first arrow) and the appearance of the surcharge on the initial customers' bills (second arrow). Much media coverage of the high bills (some twice as high as the previous month) helped call public attention to the surcharge, and following a large rainstorm, consumption reduced dramatically and

1996 Pumping – San Antonio

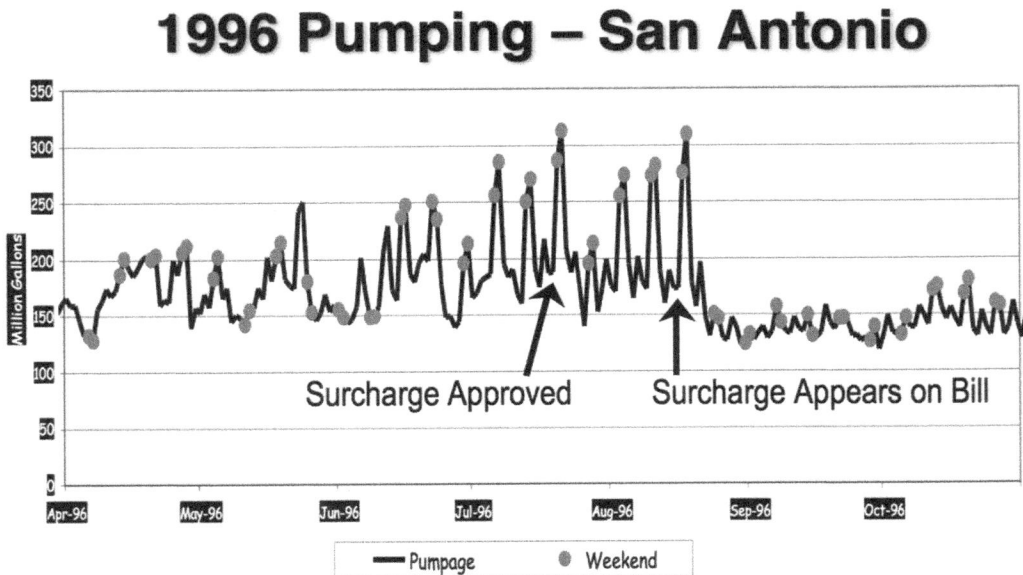

Figure 5-2   Demand response to a surcharge—San Antonio Water System

never resumed the high summer levels seen before the surcharge. More than five weeks separated the surcharge decision and the reduction in pumping.

Water suppliers may assume that they will immediately achieve the reduced levels of water use requested. In areas that have not experienced rationing before, this is unlikely. This is because other local water suppliers in the region may have differing messages that may prevent significant water use reductions. For this reason, cooperation with other local and regional water suppliers in the development of a consistent drought-related message can be very beneficial. Customer response can also be delayed because many customers learn of their water use only from their bills, many of which are mailed bimonthly.

By the time water suppliers find out that response is lagging, less water is available for the rest of the year. The likely result of this lagtime effect is that water suppliers will have to bypass the more moderate rationing levels and go to severe levels in the spring and summer. Instead of progressing from a Stage 1 rationing level to a Stage 2 level, they will have to immediately go from Stage 1 to Stage 3.

Another effect of the lag time is that water suppliers may have to draw down emergency storage and overdraft groundwater to have sufficient supplies for the later months of the year. That reduces the supply of water targeted to help meet the next year's needs.

This lagtime effect is likely to lead to unnecessary economic losses unless it is accounted for by setting significant rationing levels early in the year. It is better to have communities ration water early at levels that are uncomfortable but manageable than to wait and later have to live with extreme rationing.

It is important that the customers hear consistent messages from water suppliers in the region, particularly when they are in the same media markets. There are frequently significant differences in the supplies available to adjacent water suppliers. If customers served by one water supplier are asked to reduce their water use as much as 30 percent while their neighbors served by another water supplier are only asked to conserve 10 or 15 percent, they will question the equity of the program. This can lead to customers not meeting the reduction needs.

Once the Drought Is Over

When a drought ends, a water supplier may choose to continue some measures or incentives, to help build resilience to the next drought or to address other supply issues such as meeting peak demand during periods of high use. For example, irrigation system audits offered when the shortage is over can help educate customers and also ensure that their systems are working properly and not experiencing leaks, especially if they have been dormant for a season or two. Step 7, "Implement the Plan," ends with a discussion of other steps to consider when the drought is over.

REFERENCE

City of Santa Cruz. 2015. Urban Water Management Plan. http://www.cityofsantacruz.com/home/showdocument?id=55168 (accessed March 8, 2019).

This page intentionally blank.

Step **6**

Adopt the Plan

Each time a water shortage looms, the water shortage plan (WSP) should be reviewed and possibly updated. Each drought or other shortage situation has enough unique characteristics that a general plan cannot define all the scenarios and specific supply and demand management actions. The usefulness of a plan lies in planning the range of supply and demand management actions in advance of the situation and in defining the communication mechanisms by which decisions will be made during the event. It is important for staff to review any previous "lessons learned" or other documentation developed after a previous drought. Many staff members in charge of implementing the latest plan may not have been involved with developing and implementing the previous drought plan.

Once the updated WSP is completed, it should be formally adopted quickly—catastrophes can happen at any time. With public input and noticing requirements, adopting a plan can take one to two months. The implementation process, running concurrently with the adoption process, can usually be completed within five months. If drought conditions are imminent, the water supplier will need to treat the situation as an emergency, mobilizing the necessary people and resources to develop and implement the procedures to carry out the needed drought phase.

INVOLVE THE COMMUNITY

The key element of this step is the involvement of customers in order to create a plan that the community understands, contributes to, and supports. Public involvement will, to a large extent, determine the effectiveness and equity of the water supplier's water shortage management program. When the draft WSP is completed, it should be presented at several community forums. Making the plan available for download from the supplier's website will increase the quality of the public suggestions. The community will be more likely to reduce water use if the plan incorporates ideas suggested at the public forums. The sectors of the community that could be most affected should be included, such as industry, tourism, agriculture, the landscape community, hospitals and nursing homes, and disadvantaged communities in the service area.

Open communication is crucial. Suppliers should actively listen to, engage, and involve its customers, solicit feedback, address identified concerns, and respond to public input in a manner that is respectful, appreciative, welcoming to creative solutions, and acknowledging of each individual's contribution. Information should always contain positive action steps people can take to foster a spirit of cooperation and create an overall atmosphere that encourages the public to save water for the common good.

PREPARE A REVENUE PROGRAM

A reduction in water use will mean a revenue shortfall for most water suppliers. This is especially true with the additional costs of purchasing emergency supply and implementing demand-reduction measures. There are two common ways of balancing the costs and revenues: (1) raising water rates and (2) imposing a water shortage surcharge. If a water supplier must implement Stage 3 or Stage 4 of the WSP, rates may have to be doubled to cover fixed costs and extraordinary rationing and supplemental supply expenses. Two additional funding sources that may be available are the financial reserves in the general or water revenue fund and a designated water shortage emergency account.

Raising water rates can include an excess charge for each unit of water over the customer's allotment when rationing is in effect (Table 6-1). This helps to reinforce adherence to the allotted amounts. Forecasting the amount of revenue that will be generated, however, is more complicated when using this method. Some water suppliers have refunded excess use charges if the customer was able to repay the excess water during the rationing year. While the refunds are added work for the supplier, they build community support and trust.

Table 6-2 shows an example budget worksheet projecting the financial effects of the four stages of a water shortage plan. As water demand is reduced from normal to half of normal, it requires rate increases of 6–57 percent to keep the budget balanced. The cost of supply is dropping, but the cost of treatment increases as lower-quality sources are used. Capital projects are suspended, but the cost of the conservation program (customer assistance and rebates) and other operating expenses steadily increases.

If the water supplier simply wants to recover all of its extraordinary water shortage–related expenses and lost revenues necessary to meet fixed costs, a water shortage surcharge can be applied for the duration of the crisis. The water shortage surcharge method is easier to administer and may enable a more accurate prediction of the additional revenue that will be generated. This method is also easier for the customers to understand as a water shortage–related charge and not a disguise for a rate increase that may not end when the water shortage is over. As part of the water shortage surcharge ordinance, the termination of the surcharge should be described once the crisis is past.

The water supplier should consider the financial feasibility of funding part of the revenue shortfall from emergency reserves. It may be practical to cover as much as 50 percent of the first-year extraordinary expenses and lost revenue from such funds if they are available.

Table 6-1   Example excess use charges

Units in Excess of Allotment	Excess Use Charge per Unit
First bill, excess units	Four times normal rate
Second bill, excess units	Four times normal rate
Third consecutive bill, excess units	Ten times normal rate

Table 6-2   Water shortage plan budget worksheet

Sales	Normal	Stage 1 85% normal 6% rate increase	Stage 2 75% normal 12% rate increase	Stage 3 65% normal 26% rate increase	Stage 4 50% normal 57% rate increase
Fixed charge	$7,409,676	$7,409,676	$7,409,676	$7,409,676	$7,409,676
Quantity charge	$10,401,091	$10,643,938	$10,129,399	$9,923,566	$9,825,051
Total income	$17,810,767	$18,053,614	$17,539,075	$17,333,242	$17,234,727
Operating Expenses	**Normal**	**Stage 1**	**Stage 2**	**Stage 3**	**Stage 4**
Overhead expense	$525,500	$550,000	$575,000	$600,000	$600,000
Source of supply	$3,903,000	$3,505,170	$3,099,800	$2,647,800	$2,695,750
Product and purification	$2,000,000	$2,556,136	$2,249,840	$1,858,240	$1,716,600
Trans. and distribution	$2,500,000	$2,500,000	$2,500,000	$2,500,000	$2,500,000
Customer accounts	$850,000	$900,000	$950,000	$1,000,000	$1,000,000
General and administrative	$3,000,000	$3,300,000	$3,600,000	$3,900,000	$3,900,000
Conservation	$175,000	$300,000	$900,000	$1,200,000	$1,200,000
Depreciation	$3,600,000	$3,600,000	$3,600,000	$3,600,000	$3,600,000
Capital projects	$1,000,000	$750,000	$0	$0	$0
Total operating expense	$17,553,500	$17,961,306	$17,474,640	$17,306,040	$17,212,350
Budget balance	$257,267	$92,308	$64,435	$27,202	$22,377

Regardless of the method selected, the following actions should be included as part of the revenue program:

1. Estimate the amount of water use reduction that will be achieved and the associated lost revenue.

2. Estimate revenue needs—include funds for expensive new water supplies, increased water quality monitoring, and an extended multiyear rationing program.

3. Design a rate adjustment or water shortage surcharge that will cover the expected revenue deficit.

4. Monitor actual revenue and compare with forecast revenue; adjust water shortage surcharges as needed but not too often.

State and federal agencies offer some financial assistance to communities affected by drought. Use of such external sources of financial assistance may reduce a water supplier's revenue shortfall. However, most of these are programs of last resort, so the water supplier should be prepared to do it alone.

Formalize Cooperation With Local Agencies in the Region

Draft ordinances and interagency agreements that will be available for adoption for different levels of water shortage. They may contain various levels of mandatory restrictions and provisions that will go into effect when a state of emergency is declared by the governing body.

Regional or cooperative water supplier water shortage planning can provide a common approach to drought management among adjacent water suppliers and identification of emergency supplies and possibly provide for emergency interconnections or other joint activities. Interagency agreements confirmed in advance will speed response to an emergency and help avoid hurried decisions on matters such as price and equity. Local elected officials and city council or board members of all jurisdictions served by each supplier need to be briefed regularly and kept informed as conditions evolve.

Review and Finalize Plan

The water shortage plan should be subjected to a formal public review process. This will help minimize future objections when mandatory provisions are needed. The WSP elements and the need for them should be described in clear, concise presentations by staff to the board of directors, the public, and the media.

Several public hearings should be held on the WSP following sufficient notice by the news media. Opposition should be expected; however, ideas for beneficial changes to the plan should be welcomed. The green industry, such as landscape contractors, nurseries, etc., can mitigate economic harm during a water shortage if they are involved and are informed of potential availability of efficient irrigation systems, graywater distribution systems, and other efficient technologies. The California Landscape Contractors Association and the Irrigation Association, for example, train and certify their members as certified water managers. They can provide a list of qualified professionals that the water supplier can refer customers to when they need to improve the efficiency of their landscape water use. This is good for business for the landscape industry and good for the water supplier in terms of having confidence in the services being provided.

Contacting industry representatives ahead of time and discussing with them the opportunities and difficulties of rationing may help them understand the reasons for plan requirements and may gain their support. For example, after the city of Clinton, Okla., reached out to Mars Petcare to discuss the ongoing drought in the community, Mars Petcare voluntarily stopped irrigating, made some modifications to their cleaning processes that reduced water consumption while maintaining quality and safety standards, and fast-tracked a reuse project that would save six million gallons of water per year. Industries may also have innovative ideas that can be incorporated into the water supplier's water shortage response program.

An important internal group to coordinate with and get feedback from is the supplier's frontline customer service staff. They hear the types of questions customers are asking. They also need to know the details since they will be the ones to explain water shortage policies and procedures.

Step **7**

Implement the Plan

Following Steps 1 through 6, all of the major components of the water shortage plan (WSP) should be in place. Procedural issues, staffing needs, and budget and funding considerations should have been resolved, and the necessary ordinances and interagency agreements that underlay the plan should have been established. Step 7 provides guidance for implementing the WSP. It is recommended that the water shortage response team formed in Step 1 is the same team that implements the WSP.

This step includes three subsections. The first is a list of essential elements for successful implementation of your plan. This is followed by some considerations of each element based on the experience of this manual's authors which will assist in implementation. The final section is on communication because a successful WSP requires communicating the water supplier's goals, the actions expected from customers, and the consequences if the water shortage goals are not met.

ESSENTIAL ELEMENTS OF IMPLEMENTING A WATER SHORTAGE PLAN

Implementation of a WSP requires adequate resources to provide for the following:

1. Staff levels
2. Staff training and support
3. Office space
4. Equipment
5. Budget
6. Intra-office communication
7. Coordination with other agencies
8. Computer and billing format capabilities
9. Customer assistance

10. Customer appeals

11. Special-needs customers

12. Media contacts

13. Monitoring of actual use

Considerations

The following is an example of specific WSP implementation steps. To provide some sense of scale, the steps have been estimated for a community of 75,000 people.

1. Staff levels—first year

   a. Two full-time staff (can use reassignments), three six-month contracts, four interns.

   b. A tremendous amount of customer contact will occur when rationing or restrictions are announced, when customers receive large penalty bills, and when appeals are filed. Educate staff about the rationing program so that they are informed. The workload begins to drop off after six months.

2. Training and support

   a. Properly integrate new staff.

   b. Provide training and recognize good performance.

   c. Interns from local universities can be trained to perform water audits for those customers who exceed their allocations or ask for help. The water audit is free, and the auditors carry showerheads, faucet aerators, washers, plumbing tape, and other materials that allow them to teach customers how to make simple repairs. Water auditors also do landscape water audits, show customers how to set irrigation clocks, and demonstrate the use of soil probes.

   d. Train customer service personnel about drought restrictions so they may easily respond to customer questions.

Using "Water Cops"

A number of cities have relied on specially trained drought monitors to ensure compliance with drought restrictions. A drought monitor's (water cop's) only responsibility is to enforce the drought response measures. Temporary employees are recruited for this function, and those hired must be able to interact with the public and communicate the drought restrictions. Thorough background checks are required.

Training. Drought monitors undergo an intensive training program to prepare them to patrol the service area. Dispatchers and data entry staff participate in the training program, which will address the following.

- Dealing with irate customers
- Reading meters
- Understanding the drought restrictions
- Operating irrigation systems
- Passing driving tests
- Knowing the boundaries of their patrol area
- Understanding the data entry equipment

Documentation. Water cops use handheld data entry devices that allow them to issue tickets in the field. This system keeps track of the number of violations for each customer and enables the drought monitor to ascertain the appropriate level of violation. Drought monitors will also keep track of stops that did not result in a ticket or written warning but were merely educational in nature.

---

3. Office space

   a. Expanded office space is essential for new staff. Consider space adequate for high-volume customer traffic—a portable building, storefront, etc. Separate it from normal functions.

4. Equipment

   a. Telephones—multiple lines, cell phones, new hotline number.

   b. Computers—one per person; staff will need current information to provide correct answers to customers' questions.

   c. Cameras—staff will be taking date- and time-stamped photos; cell phones may be used.

   d. Cars—for appeals, inspections, audits, and water waste patrols.

   e. Audit kits, educational materials, water waste report forms.

5. Budget (75,000 people—$150,000 + money for rebates)

   a. The cost of staff, cars, phones, and computers.

   b. The cost of publicity, rebates, and free showerheads, nozzles, soil probes, buckets, etc.

6. Intra-agency coordination

   a. Keep everyone informed with regular email updates.

   b. Activate your water shortage response team involving billing, programming, customer service, public relations, operations and engineers, water conservation, senior management, etc.

   c. Schedule team meetings at least monthly before declaring Stage 1 and weekly afterward.

   d. Present board or policy reports summarizing each department's activities regularly during a drought.

   e. Consider tying enforcement and compliance calls directly into the water supplier's work-order system to track and process complaints more rapidly.

7. Coordination with other agencies

   a. Monthly meetings with other suppliers, city and county building and health, emergency services, county agricultural commissioner, etc.

   b. Develop phone and email group contacts for emergencies.

8. Computer and billing formal capabilities

   a. Establish flexible billing and water use history database capabilities.

   b. Adopt a flexible and informative billing format.

9. Customer assistance

   a. Establish evening and weekend hours and hotline and website information.

   b. Provide house calls, efficient landscape irrigation training, and meter reading brochure.

   c. Direct customers to the advanced metering infrastructure (AMI) customer web portal, where customers can view their hourly and daily water use data and set leak and high usage alerts.

10. Customer appeals

   a. The water shortage response team leader and appeals committee make decisions.

     b. Staff can be used either for processing appeals or helping customers make permanent efficiency changes. Design the appeal process to ask for water efficiency changes from customers.

11. Special-needs customers—hospitals, coin laundries, etc.

     a. Recognize special needs but require efficiency upgrades.

     b. Audits are always necessary.

12. Dealing with the media—publicity, customer privacy, consistency

     a. Free news coverage—the most effective tool.

     b. Keep the message consistent.

13. Monitoring of actual use

     a. Chart actual water supply and demand on a graph. A seven-day average can be used to smooth out daily fluctuations. Update the graph weekly.

     b. Compare actual demand and supply with projected demand and supply to determine if stage adjustments may be needed (Figure 7-1). Before altering the demand-reduction stage, the water supplier should consider program adjustments, such as raising the level of expenditure on public information, increasing enforcement efforts, or both. If this does not achieve the required stabilization, the reduction stage should be adjusted.

The need to have ongoing monitoring of the WSP implementation cannot be understated. The effectiveness of the individual demand-reduction measures, supply availability, and actual water use should be continually monitored. It is also important to consider impacts on water quality.

Seattle's WSP recognizes that the impacts of drought during different seasons have different effects on water quality. It found that it is essential to closely monitor water quality during droughts and particularly during warm weather. This applies to water quality in rivers as well as to the drinking water provided to customers. Water quality issues must be considered for drinking water and instream uses when supply management decisions are made. The Seattle water distribution system is designed to carry a large capacity of water during summer peak months. If demand is significantly lowered, water will not move through the

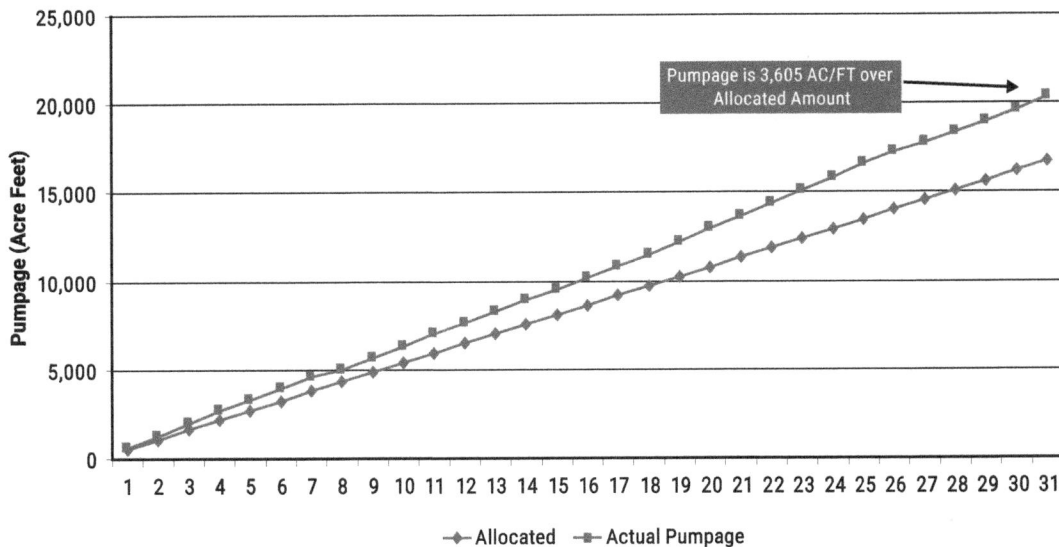

Figure 7-1   Monitoring targeted versus actual water production

system at the "design" rate. The slower-moving water, coupled with higher summer temperatures, will increase the likelihood that drinking water quality problems will arise.

The conditions leading to drought and water shortages often have unique characteristics that require an agency to modify the WSP. The WSP should be reviewed in the context of current system and forecast conditions. It should reflect current infrastructure status and capabilities. Some components may need to be revised to meet the specific conditions of a potential drought within the following 1–2 years (Figure 7-1). Begin six months before it is known for sure that the water supplier will have to implement rationing. Most water suppliers are able to forecast the likelihood of drought-related supply shortages a year before they occur.

Six to eight months is a reasonable time to develop an effective, equitable WSP that the staff can implement smoothly. The utility should have a process to periodically review and update the plan every two to three years, irrespective of an immediate need based on forecasted water supply and demand. Most states require a plan update process for water resources management every five years, which provides an opportunity to review and update the water supply triggers in the WSP.

PUBLIC INFORMATION AND MEDIA PROGRAM

Getting the public involved will require an expansion of an existing water conservation public education program. A vigorous public education program during a water shortage emergency is crucial for achieving substantial water-use reductions. The water supplier assumes a central role in publicizing the extent of the water shortage problem as well as in helping consumers conserve. Even voluntary programs have achieved significant reductions in water use where the public was well-informed and understood the need to conserve.

Aim a public education program at the following five basic groups.

1. Provide information to local decision makers regarding why certain actions are needed, why special arrangements for communication and coordination will be called for, and why both emergency funds and emergency powers may be needed.

2. Encourage governmental bodies (park and fire departments, universities, recreational facilities, and other water-dependent agencies) to provide leadership by taking timely actions to reduce demand and provide examples to the public of how they are leading by example. When drought hit Castine, Maine, in 2015, the Maine Maritime Academy asked students to conserve water during the convocation when the students returned to class. The Academy also installed low-flow toilets and showers, built a Leadership in Energy and Environmental Design (LEED)–certified building complete with its own graywater system, and even did away with lunch trays to reduce the water needed for washing. Government actions can go beyond the efforts being asked of the public and can occur quickly and at the initiation of the agencies themselves. The water supplier takes the lead and works with local elected officials and the media to promote cooperation and commitment from governments in its service area. Governments are willing to respond, especially if given technical guidance.

3. Provide detailed information to industry, schools, hotels, retailers, and other groups that are asked to comply with specific use restrictions. Also, call upon these groups to suggest alternatives to the proposed reduction measures that might achieve an equivalent level of demand reduction with potentially less

economic harm. Innovative ideas have been generated by the private sector in past droughts. At a minimum, this approach will help ensure willing participation by demonstrating a genuine interest in their perspective.

4. Provide frequent briefings to the news media to ensure timely and accurate communication. Be especially watchful for human-interest stories. Telling the media of specific instances of an individual or a group making sacrifices for the common good is a way the water supplier can show appreciation for conservation efforts and encourage others to make similar sacrifices.

5. Provide information to the public on a regular basis about the water supply situation, what actions are being proposed or taken, how those actions will mitigate supply shortages, and how well customers are meeting program goals.

When appealing to customers for water use reductions, the water supplier should act equitably, credibly, and consistently. The water supplier should demonstrate to the public that everything possible is being done to minimize the shortage. Supply options should be pursued vigorously; if new supplies are too costly or not achievable in a short time, communicate that fact. Publicity about changes in water supplier operation and maintenance practices that conserve water is helpful. Also, accurate information should be provided concerning supply status (reservoir and ground water levels), water use reductions, and other pertinent information to all water supplier personnel, especially those briefing the media or involved with public education, as well as meter readers and billing department employees.

Developing the public information campaign takes time. The 2004 Denver Drought Plan included the following media implementation plan.

Communications Plan Key in Denver Drought Plan

Once the Board has identified a specific drought stage, Denver Water's Community Relations staff will develop an appropriate communications plan based on the elements specified in the Outline for a Drought Communication Plan.

February
- Select advertising agency to assist staff with mass media advertising campaign.
- Announce "Spring Watch" (voluntary ban on lawn watering), subject to Board decision.
- Promote relevant news story topics to media and respond to media inquiries.

March
- Begin developing message-of-the-week program.
- Reinforce voluntary ban on lawn watering.
- Board contracts with ad agency.
- Community Relations staff and agency begin work on campaign.
- Hold public meeting about surcharges.
- Promote relevant news story topics to media and respond to media inquiries.

April
- Begin disseminating message of the week.
- Reinforce voluntary ban on lawn watering.
- Promote relevant news story topics to media and respond to media inquiries.
- Board determines drought stage and corresponding drought response measures.
- Post drought stage and response measures on website.
- Intensify media relations.
- Community Relations staff prepares and mails notices to all customers.
- Board approves ad campaign.

Communications Plan Key in Denver Drought Plan
May   • Disseminate message of the week.   • Ad campaign begins.   • Mail notices of drought response measures to all customers.   • Promote relevant news story topics to media and respond to media inquiries.

In dealing with the media, one person should speak for the water supplier—preferably the water shortage response team leader. Media inquiries should be replied to immediately to maintain communication links and to prevent media representatives from seeking alternative information sources that may be less informed. Good communication provides opportunities for a water supplier to tell its story and ensures that knowledgeable people will be called on to speak on the issues.

Before developing water shortage-related public information strategies, there are several important issues to address about the program focus and content. First, the water supplier should emphasize that the water supply situation is unpredictable and may change from month to month. No one can be certain when the situation will improve. Even if precipitation increases, the effect on the water supply may not be immediate. The water supplier needs to proceed cautiously by starting demand-reduction measures early and should avoid relaxing any measures too soon. Also, customers need to realize that the drought impact is not uniform across a state or region and that the problem will be more severe in some areas and less severe in others.

Some classes of customers may carry the burden of coping with the water shortage more than others. Some groups with high potential for reduction may be asked to reduce water use more than others but avoid discrimination within a class of customers. Landscape irrigation may have to be curtailed. Conversely, it may be decided to minimize water reductions to commercial-industrial customers to preserve as many jobs as possible. The reasoning behind these or any mandatory curtailment of supplies should be carefully communicated.

The public should be made aware of the impact of the water shortage on water supplier costs as early as possible. Reduced water sales will obviously reduce revenue. Most water suppliers have fixed costs on approximately 75–80 percent of their total budget, and the public needs to know this. There may be significant additional costs incurred for purchasing water, conservation programs, emergency pumps, pipes, other equipment, increased water quality testing, and other water shortage–related activities. These costs will be borne by the customers.

Finally, the water supplier should avoid being placed in an adversarial position. The focus should be on the emergency at hand without blame implied toward the water supplier's management or a customer class.

It is important to tailor the public information program to the type of community served. For large decentralized areas, methods that allow the water supplier to reach many customers relatively inexpensively, such as websites, email, direct mail, bill inserts, and media advertisements, are appropriate. Smaller, close-knit communities with central business districts may also be well served by a central information center.

Public information programs provide long-term benefits by increasing the customers' understanding of their water use and of the water supplier's operations. Such an understanding will be useful in generating public support for future efforts regarding rate increases or new efficiency and supply projects.

Tampa Bay Water Highlights How Wholesalers Can Work with Retail Utility Customers During Drought

Tampa Bay Water (TBW) is a wholesale water provider that provides services to 2.5 million customers through six member-government water agencies. Its Water Shortage Mitigation Plan provides TBW with a strategy for identifying and responding to shortages caused by drought conditions. The plan, updated most recently in 2017, includes four stages of water shortage conditions based on hydrologic triggers. The plan defines the triggers for entering and exiting each of the shortage stages. The triggers are based on current measurements of regional rainfall deficits, streamflow, and actual and predicted (3-month) reservoir levels. Likewise, for each stage the plan describes the supply management actions that will be taken by TBW as well as demand-management actions that will be taken by both TBW and the member agencies.

Communication is a critical component of the plan because many of the demand actions depend on cooperation from customers. The plan describes processes for keeping key stakeholders informed, managing media communications, and providing outreach to customers about efforts to reduce demand. The plan highlights the importance of developing and delivering consistent messaging to customers throughout the region during a water shortage.

"While preferred delivery methods among members may vary, the uniformity of the key message and timing of delivery occurring throughout the region during a water shortage can be effectively managed with proper planning and coordination."

The plan, available at https://www.tampabaywater.org/documents/conservation/Water-Shortage-Mitigation-Plan-2017.pdf, also provides a useful flowchart (Figure 7-2) that summarizes the implementation of actions associated with the plan.

When undertaking any public information effort, it is crucial that the information be accurate and consistent and that requested use reductions be commensurate with the seriousness of the situation. In other words, the customer must understand what the trigger conditions are, what the consequences of the different stages of drought are, and how the emergency measures will help relieve or minimize the problem. The example from Tampa Bay Water illustrates the importance of communication in overall water shortage plan implementation as well as in wholesaler and retailer coordination.

DROUGHT RECOVERY AND WATER SHORTAGE PLAN TERMINATION

Droughts vary in their intensity and duration. A water shortage may last only a few months or extend over a prolonged period of several years. Drought conditions will eventually ease, allowing suppliers to downgrade or terminate the plan.

A water shortage ends when water supply conditions, including stream flows, reservoir levels, and groundwater storage improve to the point that a water system is capable of supporting unrestricted water demand for a sustained period of time. In some cases, the end of a drought is plainly evident. In others, differences may arise among scientists, resource agencies, regulators, and other interested parties about whether conditions have improved sufficiently to safely lift all water shortage restrictions and rules. Also, not every supplier in the same region will recover at the same rate. Some areas may improve quickly. For others, drought conditions can drag on due to differences in sources and type of water supply. Each water supplier, therefore, should consider developing specific water supply triggers or criteria for its system that, once met, signal the end of a drought.

As with other emergency preparedness activities, WSPs should be periodically reviewed and updated. It is appropriate after a drought ends for the water shortage response team to reflect on actions taken and their effectiveness. Planning activities should

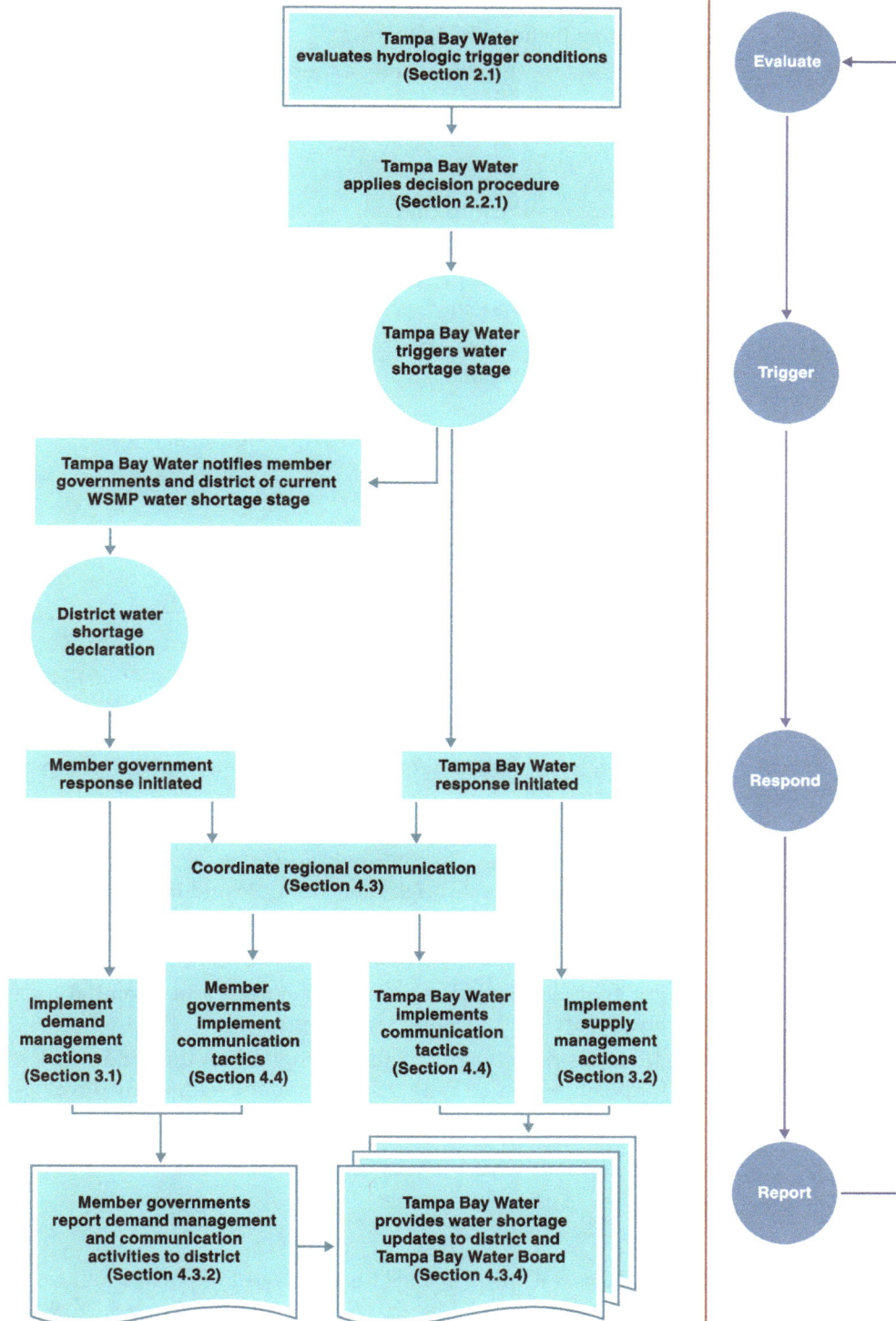

WSMP: Water shortage mitigation plan
Adapted from Tampa Bay Water

Figure 7-2   Example flow chart for regional cooperation

continue even after the administrative apparatus created to respond to drought has wound down. Such activities may include the following.

- Publicize gratitude for the community's cooperation
- Restore water utility operations, organization, and services to pre-event levels
- Document the event and response, and compile applicable records for future reference
  - Collect cost accounting information, assess revenue losses and financial impact, and review deferred projects or programs
  - Debrief staff to review effectiveness of actions, to identify the lessons learned, and to enhance response and recovery efforts in the future
  - Complete a detailed evaluation of affected facilities and services to prepare an "after action" report, including lessons learned and recommended improvements
- Continue to maintain liaisons as needed with external agencies
- Update the WSP as needed

CONCLUSIONS

This manual presents a process to develop, implement, and revise a water supplier's water shortage plan to address the challenges associated with drought. Throughout this process, it is helpful to remember that such challenges often present opportunities. The Rural Water Users Association of Utah observed the following positive impacts of an extended drought.

- State legislation was passed requiring water conservation planning by cities, towns, and districts.
- Cities were motivated to prepare drought response plans, evaluate tiered water rates, and become conscious of their own water use and the messages they were sending to their customers through their example and through the media.
- More cities were motivated to complete their water conservation plans and to adopt a time of day watering ordinance.
- The drought sparked more discussion of the need and potential for water reuse planning and future funding sources.

While drought cycles pose a significant challenge to communities, they can also create opportunities for positive change in overall water management. By developing a robust and flexible water shortage plan, a water supplier has made an important investment in the reliability and resiliency of their supplies.

REFERENCE

Tampa Bay Water. 2017. *Water Shortage Mitigation Plan*. https://www.tampabaywater.org/documents/conservation/Water-Shortage-Mitigation-Plan-2017.pdf (accessed March 8, 2019).

Appendix **A**

Water Shortage Planning Checklist

First Steps

- Designate water shortage response team leader
- Designate team members from each department or division
- Set priorities
- Identify potential supplemental supply sources
- Identify potential interconnections
- Identify regional suppliers for potential cooperative actions
- Determine sources for long- and short-term weather and supply predictions
- Establish a community advisory committee

Supply

- Quantify worst-case supply (minimum) for next five or more years
  - Local surface
  - Wholesale
  - Groundwater
  - Recycled
  - Other

Water Quality

- Project water quality changes by source
- Identify water treatment devices necessary to use on degraded quality sources
- Identify low-quality water sources and develop plan for blending

Demand

- Quantify worst-case demand (maximum) by season for next five or more years
  - Single family
  - Multifamily
  - Commercial
  - Industrial
  - Institutional
  - Landscape
  - Recycled
  - Agricultural
  - Wholesale
  - New connections

Supply and Demand Balance

- Quantify yearly shortage for next five or more years
  - 2019
  - 2020
  - 2021
  - 2022
  - 2023

Increase Supply

- Project possible supplemental supplies and carryover
- Schedule well driller for new or rehabilitated wells
- Plan to increase supplier efficiency
  - Meters
  - System losses
  - System pressure
  - System flushing
  - Supplier landscaping

Decrease Demand

- Determine health and safety minimum supply
- Plan Stage 1—public relations campaign and recommend customer actions
- Adopt and publicize water waste ordinance and time-of-day irrigation restrictions
- Make available nonpotable water stations for nonpotable uses
- Review pricing structure and rates by stage
- Select water allocation method by customer class and stage

- Adopt restriction-enforcement rules and penalties
- Select stage and customer class demand-reduction programs to help customers
- Plan for catastrophes with cascading failures—50 percent supply shortage or more

Complete Draft Water Shortage Plan (WSP)

- Establish stage triggers based on priorities and quantifiable supply availability by source
- Include carefully crafted flexibility to triggers
- Identify lag time and seasonal issues related to each reduction program
- Establish structure and impacts of limited-number-of-days irrigation programs
- Develop revenue plan to balance budget by stage
- Develop customer appeal procedure
- Establish monitoring program to track water production and use
- Provide for orderly and adaptive steps to move to lower stages when the drought recedes

Community Involvement

- Complete Draft Plan
- Provide Draft Plan to community
- Contact significantly impacted customers (agriculture, green industry, tourist industry, etc.) and request input
- Contact local suppliers and government agencies and request input
- Hold at least three public meetings to receive comments on Draft Plan
- Incorporate useful community suggestions into Draft Plan
- Adopt the final WSP

Supplier Capabilities and Resources

- Establish required computer capabilities for billing, data tracking, and customer support
- Identify required changes to existing computer systems
- Make required computer system changes and test thoroughly
- Prepare customer information brochures
  - Meter reading
  - Leak detection
  - Plumbing hardware recommendations and rebate programs
  - Customer assistance programs offered by supplier staff
- Identify needed new full-time and part-time contract staff
- Procure space for additional staff and increased customer visits
- Develop media contacts
- Identify and purchase water conservation devices for distribution to customers
- Develop training program for staff
- Develop training programs for affected businesses
- Establish water waste and information hotline

This page intentionally blank.

Appendix **B**

Additional Sources of Weather and Climate Information

National Integrated Drought Information System—Home to drought.gov and a one-stop shop for all information drought-related	https://www.drought.gov
National Weather Service—Main site for current weather conditions	https://www.weather.gov/
USGS Water Resources—Provides access to historic and real-time stream flow and water quality data	https://www.usgs.gov/science/mission-areas/water-resources
USDA—The Natural Resources Conservation Service (NRCS) offers reports on snowpack and water supply conditions as well as state water supply outlook reports for the western states and Alaska	https://www.wcc.nrcs.usda.gov/ and https://www.wcc.nrcs.usda.gov/state_outlook_reports.htm
Several states are also developing platforms that support drought monitoring, climate conditions, and water supplies:	
Arizona reports on current drought status using a weekly map of drought conditions and a monthly drought status summary.	https://new.azwater.gov/drought/drought-status
Georgia provides drought indicator reports that are released with a frequency that is dependent upon how recently there has been a drought.	https://epd.georgia.gov/water-conservation
California provides a *California Water Supply Outlook* from the DWR California Data Exchange Center which complements the USGS information in that it presents current California data on hydrologic conditions such as snowpack, runoff, and reservoir storage and is published twice monthly. The DWR also maintains a portal page for drought with other information.	http://cdec.water.ca.gov/ and https://www.water.ca.gov/Water-Basics/Drought
Texas provides real-time information on the status of supplies in the state, including reservoirs.	https://waterdatafortexas.org/reservoirs/statewide

This page intentionally blank.

Glossary

Alliance for Water Efficiency (AWE)—A nonprofit membership organization promoting the efficient and sustainable use of water. www.allianceforwaterefficiency.org

Allocation—An amount of water apportioned to each customer or customer class. Excess use can be penalized by fines, rate surcharges, or other restrictions. Methods for determining allocation amounts are discussed in Step 5 of this manual.

American Water Works Association (AWWA)—A professional organization serving the drinking water supply profession.

Class—Customers having similar characteristics (commercial, single-family residential, etc.) grouped together for billing or program purposes.

Conjunctive uses—Integrated management and use of two or more water resources, such as an aquifer and a surface water body.

Consumptive use—The quantity of water that is not available for immediate reuse because it has been evaporated, transpired, or incorporated into products, plant tissue, or animal tissue. Also referred to as "water consumption."

Demand management—Measures, practices, or incentives deployed by utilities to change the pattern of demand for a service by its customers or slow the rate of growth for that service.

Drought—A period of lower than normal precipitation. Can also be defined by a deficit in soil moisture, or a shortage in water supplies caused by low river flows, low reservoir levels, or increases in depth to groundwater.

Enforcement—Actions taken by a water agency or local municipality to ensure that mandatory demand management activities are taken by all customers. Fines or bill surcharges are typical tools used to compel violators to comply.

Evapotranspiration (ET)—A measure of the combined amount of water transpired by plants and evaporated from plant and soil surfaces. Typically, ET is not measured directly but is estimated from climatic data measured by a weather station (called Reference Evapotranspiration, ETo) multiplied by a plant factor.

Graywater—Domestic wastewater composed of wash water from kitchen sinks, bathroom sinks and tubs, clothes washers, and laundry tubs that can be used for nonpotable purposes such as irrigation.

High water-use landscape—A landscape made up of plants, turf, and features that requires 80 percent or more of the reference evapotranspiration to maintain optimal appearance.

Informative billing—System of providing water utility customers with useful information on the relationship between the amount of water they use and the cost associated with that use. Examples of the information include the utility's rate structure, amount of water used in the current month, amount of water used in the previous month, amount of water used in the same month of the previous year, information on the average usage of all customers in the same customer class, seasonal rates and applicable months, drought rates, information on conserving water, or other information deemed appropriate by the utility.

Measure—An action, behavioral change, device, technology, or improved design or process implemented to reduce water loss, waste, or use. It should be noted that the value and cost-effectiveness of a water efficiency measure must be evaluated in relation to its effects on the use and cost of other natural resources (e.g., energy and/or chemicals).

Multifamily unit—A residential dwelling consisting of multiple housing units contained within one building (e.g., an apartment building) or several buildings within a complex.

Per capita use—The amount of water used by one person during one 24-hr period. Typically expressed as gallons per capita per day (gpcd).

Reclaimed water—Municipal wastewater effluent that is given additional treatment and distributed for reuse in certain applications. Also referred to as *recycled water*.

Recycled water—A type of reuse water usually run repeatedly through a closed system; sometimes used to describe reclaimed water.

Reference evapotranspiration (ETo)—The estimate of the evapotranspiration of a broad expanse of well-watered, 4-to-7 in.-tall cool-season grass. It is computed from climatic factors such as solar radiation, wind speed, air temperature, and relative humidity.

Retrofit—(1) Replacement of existing water using fixtures or appliances with new and more efficient ones. (2) Replacement of parts for a fixture or appliance to make the device more efficient.

Reuse—Use of treated municipal wastewater effluent for specific, direct, beneficial uses. See *reclaimed water*. Also used to describe water that is captured onsite and used in a new application.

Safe yield—The amount of water that can be withdrawn annually without negative effects. Safe yield can be controversial in areas where ground subsidence, overlapping cones of depression, and the impacts of source water reduction are detrimental to the environment.

Single-family unit—A residential dwelling unit built with the intent of being occupied by one family. It may be detached or attached (e.g., townhouses).

Soil amendment—Organic and inorganic materials added to soils to improve their texture, nutrients, moisture-holding capacity, and infiltration rates.

Surcharge—A special charge included on a water bill to recover costs associated with a particular activity, facility, or use or to convey a message about water prices to customers. Used by some water agencies to send a price signal during water shortages.

Transpiration—The passing of water through living plant membranes into the atmosphere.

Water conservation—Any beneficial reduction in water use or in water losses. Activities designed to reduce the demand for water, improve efficiency in use, and reduce losses and waste of water.

Water efficiency measure—A specific tool or practice that results in more efficient water use and thus reduces water demand.

Water losses—Metered source water less metered and authorized unmetered water use.

Water Research Foundation—A nonprofit organization that sponsors research for the drinking water supply profession.

Water shortage—Any period in which a water supplier is unable to meet the typical unconstrained demand. It can be caused by a drought but could also be a result of equipment failures or contamination of water supplies.

Water surcharge—Imposition of a higher rate on excessive water use.

Water waste prohibition—A water conservation measure where the water supplier develops and enforces an ordinance or rule prohibiting certain excessive uses of water.

Xeriscape—A trademarked term denoting landscaping that involves the selection, placement, and care of low-water-use plants. Xeriscape is based on seven principles: proper planning and design, soil analysis and improvement, practical turf areas, appropriate plant selection, efficient irrigation, mulching, and appropriate maintenance.

Index

Note: *f.* indicates figure; *t.* indicates table

AWWA Manuals

M65,   *On-Site Generation of Hypochlorite, #30065*

M66,   *Cylinder and Vane Actuators and Controls—Design and Installation, #30066*

M68,   *Water Quality in Distribution Systems, #30068*

M69,   *Inland Desalination and Concentrate Management, #30069*

M77,   *Condition Assessment of Water Mains, #30077*